Understanding Sustainable Development

Understanding Sustainable Development

Miles Glover

www.callistoreference.com

Callisto Reference,
118-35 Queens Blvd., Suite 400,
Forest Hills, NY 11375, USA

Visit us on the World Wide Web at:
www.callistoreference.com

ISBN: 978-1-64116-625-6 (Hardback)

Cataloging-in-Publication Data

Understanding sustainable development / Miles Glover.
    p. cm.
Includes bibliographical references and index.
ISBN 978-1-64116-625-6
1. Sustainable development. 2. Economic development--Environmental aspects.
I. Glover, Miles.
HC79.E5 U53 2022
338.927--dc23

TABLE OF CONTENTS

The purpose of this book is to help students understand the fundamental concepts of this discipline. It is designed to motivate students to learn and prosper. I am grateful for the support of my colleagues. I would also like to acknowledge the encouragement of my family.

Sustainable development refers to the organizing principle to meet the human development goals while concurrently sustaining the ability of natural systems to provide ecosystem services and natural resources. The three main spheres where it is applied are environment, economics and society. Environmental sustainability deals with the natural environment and how it can remain diverse and productive. In terms of economics, sustainability involves ecological economics, environmental economics and sustainable energy. Environmental politics and environmental governance are a few focus areas related to social sustainability. This book is a compilation of chapters that discuss the most vital concepts and emerging trends in the field of sustainable development. Different approaches, evaluations, methodologies and advanced studies have been included herein. As this field is emerging at a rapid pace, the contents of this book will help the readers understand the modern concepts and applications of the subject.

A foreword for all the chapters is provided below:

Chapter – Introduction

Sustainable development is defined as the development that meets the needs of the present without compromising the ability of future generations to meet their own needs. The topics elaborated in this chapter will help in gaining a better perspective about sustainable development and its goals.

Chapter – Environment and Sustainable Development

The responsible interaction with the environment for the purpose of avoiding the depletion or degradation of natural resources and allowing for the long-term environmental quality is known as environmental sustainability. This chapter has been carefully written to provide an easy understanding of the different environmental issues as well as climate change mitigation.

Chapter – Sustainable Development: Theories and Perspectives

A few perspectives and theories related to sustainable development are the tragedy of the commons, eco-development and Gandhian model. These topics explained in this chapter will help in gaining a better perspective about these theories and perspectives of sustainable development.

Chapter – Strategies for Sustainable Development

There are various kinds of strategies used for sustainable development. A few of them are sustainable agriculture, sustainable transport and sustainable forest management. This chapter closely examines these key strategies to provide an extensive understanding of the subject.

Chapter – Sustainable Technologies

The technologies which focus on reducing waste while maintaining efficiency as well as improving the social and environmental footprint during the production cycle are known as sustainable technologies. The diverse aspects of sustainable technologies and sustainable energy have been thoroughly discussed in this chapter.

Chapter – Sustainable Waste Management

Sustainable waste management reduces the amount of natural resources consumed and ensures that any materials that are taken from nature are reused as many times as possible. This chapter has been carefully written to provide an easy understanding of the varied facets of sustainable waste management.

Miles Glover

Introduction

Sustainable development is defined as the development that meets the needs of the present without compromising the ability of future generations to meet their own needs. The topics elaborated in this chapter will help in gaining a better perspective about sustainable development and its goals.

SUSTAINABLE DEVELOPMENT

Sustainable Development is a concept that at its core is revolutionary, yet unfortunately incredibly difficult to pragmatically define. The history behind sustainable development is one that does not stretch far. Tensions that can be found within the concept of sustainable development are numerous, ranging from its ambiguous and vague definition, to the failure of attaining a universal pragmatic and operational framework. The great challenge that lies ahead with sustainable development is not only the need to educate it to the people, but to first define it in a way people will understand it.

The concept of sustainable development is one that arguably is multi-disciplinary, complex, and systematic, yet defining the concept is without a doubt a great task. Sustainable development was a term first coined in 1980, when the intent of the concept was merely basic. It was in the World Conservation Strategy, a union between three prominent environmental non-governmental organizations IUCN, WWF, and UNEP, where sustainable development took on the meaning of 'conserving the earth's natural resources'. What the World Conservation Strategy had realized is that with the world's economic growth, came the near-sighted exploitation of the world's natural resources. The original, and sole, intent of the World Conservation Strategy was to bring nations together to stop the exploitation of natural resources, which in turn was negatively affecting the environment. Sustainable development was thus merely seen on quite a basic level, at the time of its coinage. Not even a decade later, did the definition take on much more of a multi-disciplinary approach. In 1987 the white paper, named Our Common Future, was published by the World Commission on Environment and Development (WCED). The document set the loose foundation of sustainable development with a widely quoted definition, which states "sustainable development is development that meets the needs of the present without compromising the ability of future generations to meet their own needs". The document, also frequently known as the Brundtland Report, has "since been taken up by almost every international institution, agency and NGO". The Brundtland Report became the first document to support sustainable development as a multi-disciplinary field, as it explained that the economy, society, and the environment were key to sustainable development. In 1992 the United Nations Conference on Environment and Development devised a program entitled Agenda 21, which allegedly "is the blueprint for sustainability in the 21st century". It is a framework that nations and government strictly can adhere to. Nations that gave their consent to accepting Agenda 21 are monitored by the United Nations Commission on Sustainable Development (CSD), who "is responsible for reviewing progress in the implementation of Agenda 21 and the Rio Declaration on Environment and Development; as well as providing policy guidance to follow up the Johannesburg Plan of Implementation (JPOI) at the local, national, regional and international levels". Both the Agenda 21 as well as the Brundtland Report have proven to be widely used frameworks that nations, agencies, and organizations use in modern times, yet even though they are accepted, a clear definition still is not evident.

The greatest problem with the concept of sustainable development is the sheer amount of definitions that are available. Unfortunately, a clear definition of sustainable development has still not been devised. Although the Brundtland Report's definition is widely-quoted, one can see exactly where it fails. When relooking at the definition "sustainable development is development that meets the needs of the present without compromising the ability of future generations to meet their own needs", one can spot two clear issues. Firstly, the 'needs' are not defined. There is a major difference between the needs of a person living in a Third-World country, as opposed to the needs of a person living in Western Europe. Secondly, the definition does not offer any type of time frame, as 'generations' can only be vaguely interpreted. The incredible amount of definitions available for sustainable development, thus make it a topic that the common man does not wish to pursue. An interesting way of looking at the issues at hand with sustainable development is too look at the following quote, which states "a combination of uncertainty about what to do, and a feeling of guilt about what is not being done, means that many people seem afraid to expose what they feel is their lack of understanding of sustainable development. Therefore, it is often easier to pretend that it does not need to be addressed". It is quite clear that because of sustainable development's uncertain definition people, nations, and governments seem to act on it in varying ways. Yet, for those that have agreed on a definition, the problem of implementation arises.

Countries that have made a conscious effort to understand sustainable development and are willing to make changes, both nationally as well as internationally, face the problem of implementing sustainable development-geared policies. As with Agenda 21, the document that gave a framework to countries for sustainable development, there is no 'enforcer' of the document. That is to say, countries may claim to understand sustainable development, agree to make changes to their policies, but nobody actually enforces them to do so. As it was stated in the Brundtland report, "our inability to promote the common interest in sustainable development is often a product of the relative neglect of economic and social justice within and amongst nations" which sheds light on yet another paramount issue in sustainable development, that stretches far beyond simply sustainable development. The problem with the enforcement of sustainable development policies is that countries that are in the greatest need of them have no reason to adhere to the policies. As the North-South dialogue still clearly exists, the North essentially is telling the South the proper way to develop. This seemingly condescending method of saying "you must do it this way", not only contributes to the hegemonic nature of the West, but simply further will concentrate the interests between both the North and South. As mentioned in an Organisation for Economic Co-Operation and Development (OECD) report, Unfortunately, the age of individuals is still strongly evident in the world today. It is however in the nature of sustainable development to distinctly outline the issues that not only a nation has with its policies, but also the world has with its tensions. Sustainable development is thus a concept that is revolutionary, yet limited in the scope of its beholder.

Sustainable development has the potential of being a groundbreaking concept that can revolutionize the way nations act on a national level, and more so on an international level. Unfortunately though, due to its multi-disciplinary nature, ideal-oriented goals, and flexible interpretations, a clear definition of the concept has yet to be found. Moreover, its vague interpretation and ambiguity further add to the tensions found within this concept, as any country could state they are following sustainable development policies. The challenge ahead is firstly to devise a concise definition, and secondly to pragmatically be able to apply it to any nation across the world. Due to sustainable development's potential of further dividing the North and South dialogue, every country must make the necessary changes to their own policies in order to provide for a cleaner, safer, and more efficient environment, economy, and society. What sustainable development essentially calls for is transparency as well as cooperative nations that are willing to work together for the betterment of the world. It is perhaps for this sole reason that sustainable development is quite difficult to define, because every nation has a different view of what a better world is.

Two fundamental elements of the concept of sustainable development, i.e. development and sustainability, preceded the creation of the concept itself. According to Sharpley, development and sustainability could be in the juxtaposition, where both could have possible counterproductive effects, while neoclassical economists emphasize that there is no contradiction between sustainability and development also suggests how there is no development without sustainability or sustainability without development. The notion of development is related to the past western concept of imperialism and colonialism, and in that period it implied infrastructure development, political power, and economic policy, serving imperialists as an excellent tool for marginalization and diminishing the power of certain countries. Certain authors link the meaning of development to economic development and the term "underdeveloped areas" (later called "Third World Countries"), which US President Harry Truman introduced in the mid-20th century, signifying areas with the significantly lower standard of living than developed areas.

Classical theories of development consider development within the framework of economic growth and development. According to these theories, development is a synonym for the economic growth that every state in a particular stage has to undergo, driven by the transformation of traditional agriculture into modern industrialized production of various products and services, i.e. shifting from the traditional society to the stage of maturity and high consumption. These theories consider developing countries as countries limited by the poor allocation of the resources emerging as a result of the firm hand of government and corruption, inefficient and insufficient economic initiatives, but also political, institutional and economic austerity, whereby being captured in dependence and domination of developed wealthy states. According to several neoliberal and modern development theories established over the past 60 years and the contemporary understanding, development is a process whose output aims to improve the quality of life and increase the self-sufficient capacity of economies that are technically more complex and depend on global integration. Fundamental purpose of this process is a creation of stimulating environment in which people will enjoy and have long, healthy and creative life. Romer's new or endogenous growth theory suggests that economic growth is a result of the internal state or corporate system, and the crucial role in economic growth is knowledge and ideas. The endogenous growth theory model consists of four basic factors: 1) capital measured in units of consumer goods, 2) labour involving the individual skills, 3) human capital comprising education, learning, development and individual training, and 4) technological development. In accordance with this model, if countries want to stimulate economic growth, they have to encourage investment in research and development and the accumulation of human capital, considering that appropriate level of the state capital stock is the key of economic growth.

In the literature different taxonomies of the meaning of the term development are found, and most often the following meanings are emphasized: 1) development as structural transformation, 2) human development, 3) development of democracy and governance, and 4) development as environmental sustainability describes development as a process of targeted change, which includes goals and resources to achieve these goals. According to Thomas, development involves the positive changes that society has experienced throughout history, and still experiences, while Sharpley's development outlines the plans, policies, programmes and activities undertaken by certain institutions, governments and other governmental and non-governmental organizations. Accordingly, the most acknowledged development indicator is the Human Development Index (HDI) which integrates different categories of socio-cultural, economic, ecological and political development of particular areas. The term sustainability literally means "a capacity to maintain some entity, outcome, or process over time" and carrying out activities that do not exhaust the resources on which that capacity depends. Since this is a general understanding of sustainability, this meaning can be placed analogously to all human activities and business processes. Thus, according to the general definition, each activity can be carried out in volume and variations without leading to self-destruction, but allowing a long-term repetition and renewal. However, Shiva points out that the general understanding of sustainability is dangerous because it does not respect the environmental limits and the need for adapting human activities to the sustainability of natural systems. Natural systems enable people to live and support the outcomes of human activities, therefore sustainability can hardly be considered without an ecological aspect. Accordingly, ecological sustainability has become a fundamental framework for considering socio-cultural and economic sustainability, but also a subject of arguing in the concept of sustainable development.

In the 18th century economic theoreticians such as Adam Smith pointed out issues of development, in the 19th century Karl Marx and classical economists Malthus, Ricardo and Mill also argued about certain elements of sustainable development, while later neoclassical economic theory emphasized the importance of pure air and water and renewable resources (fossil fuels, ores) as well as the need for government intervention in the case of externalities and public goods. Previous periods, and even the following century, saw the dominance of the economic doctrine with focus on human as a ruler of natural resources. The term sustainable development was originally introduced in the field of forestry, and it included measures of afforestation and harvesting of interconnected forests which should not undermine the biological renewal of forests. This term was firstly mentioned in the Nature Conservation and Natural Resources Strategy of the International Union for Conservation of Nature published in 1980. Although initially sustainable development primarily viewed an ecological perspective, soon it spread to social and economic aspects of study.

Development based on economic growth remained until the 1970s when it was obvious that consumerism and economic growth put pressure on environment with the consequences of polluted and inadequate living space, poverty and illness. At the same time, the exploitation of natural resources, in particular the stock of raw materials and fossil fuels, has led to deliberation of the needs of future generations and created a prerequisite for defining the attitude of long-term and rational use of limited natural resources. The imbalance between human development and ecological limits has pointed to the growing environmental problems and possible consequences with disastrous proportions. Črnjar

& Črnjar summed up the basic causes of environmental pollution: 1) anthropogenic causes of environmental pollution (economic growth, technical and technological development, industrial development, development of traffic and transport infrastructure, population growth and urbanization and mass tourism), 2) natural causes of environmental pollution (soil erosion, floods, earthquakes, volcano eruptions, fires, droughts and winds) and 3) other causes of environmental pollution (wars, insufficient ecological consciousness, imbalance between development and natural ecosystems and limited scientific, material, organizational and technological opportunities of society). The consequences of these factors – seen in various ecological problems, ecosystem disturbances, global climate change, natural catastrophes, hunger and poverty, and many other negative consequences – have been warning about the sustainability of the planet.

Aspiration of developed countries to improve the socio-economic and ecological situation of developing and undeveloped countries gathered scientists, economists and humanists from ten countries in Rome in 1968 to discuss the current problems and future challenges of humankind (limited natural resources, population growth, economic development, ecological problems, etc.). Grouped as an independent global organization called the Roman Club, these scientists have published two significant editions – Limits of Growth in 1972 and Mankind at the Turning Point in 1974, containing the results of their research and appealing the world to change the behaviour toward the planet, while in the first edition the term sustainability was clarified in the framework of the contemporary concept of sustainable development. The Roman club warned that excessive industrialization and economic development would soon cross the ecological boundaries. In 1971 Nicholas Georgescu-Roegen published The Entropy Law and the Economic Process, similarly warning about the dangers of economic development and marking the beginning of the ecological economics and environmental economics.

Different organizations and institutions participated in the creation of the concept of sustainable development. The most significant is the United Nations (UN), founded in 1945 with headquarters in New York, which nowadays includes more than 190 member states. Its main goals include: maintaining the peace and security in the world, promoting sustainable development, protecting the human rights and fundamental freedoms, promoting the international law, suppressing the poverty and promoting the mutual tolerance and cooperation. Since its establishment, UN has been active in the field of sustainable development by organizing numerous conferences, taking actions and publishing various publications aimed to achieve the goals of sustainable development and the Millennium Development Goals (MDGs). A total of 33 programmes, funds, specialized agencies and affiliated organizations are active within the United Nations, while some of them play a significant role in the creation and implementation of the concept of sustainable development. The United Nations Division for Sustainable Development (UNDSD) has also been established to promote and coordinate the implementation of sustainable development, particularly in the field of intergenerational and international co-operation. The Division also serves as a support to policy management and management of sustainable development, and especially as a communication platform for knowledge and data dissemination. Along with this, the UN has established a Global Network of Sustainable Development (GNSD) geared to achieve the Millennium Development Goals.

Since the introduction of the concept, many international conferences, congresses, summits and meetings have been held, resulting in various declarations, reports, resolutions, conventions and agreements and dealing with the environmental problems.

Among the various activities, three key events set the fundaments and principles of sustainable development. According to them, the history of the concept of sustainable development is divided into three periods. The first period covers the period from economic theories, where certain theorists recognized the boundaries of development and environmental requirements, through the activities of the Roman Club, which warned on the negative consequences of economic development, to the First United Nations Conference on the Human Environment held in Stockholm in 1972. This conference marked the introduction of the concept of sustainable development, and although it did not fully associate environmental problems with development, it stressed the need for changes in economic development policy. In the report published after the conference, the necessity of balance between economic development and environment was proclaimed and 28 principles were set aimed to preserve environment and reduce poverty. Within the action plan, 109 recommendations (socioeconomic, political and educational) were given for quality environmental management, and finally, after the conference, resolution on institutional and financial agreements was signed between the states.

Years after the Stockholm conference represent the second period of the concept of sustainable development. The terms such as development and environment, development without destruction and development in accordance with the environment were increasingly used in publications, while the term eco-development was first described in edition of the United Nations Environment Programme (UNEP) published in 1978. In 1980, International Union for Conservation of Nature (IUCN) set an idea of linking economics and the environment through the concept of sustainable development. A few years later, more precisely in 1983, the United Nations World Commission on Environment and Development (WCED) was established to develop a global change programme. This programme was aimed to raise awareness and concern about the negative impact of socio-economic development on the environment and natural resources as well as provision of perspectives of a long-term and sustainable development in accordance with the environmental protection and conservation. After several years of work, in 1987 the Commission of 19 delegates from 18 countries, led by Gro Harlem Brundtland (the then Norwegian Prime Minister), published a report Our Common Future, better known as the Brundtland Report, where the concept of sustainable development was introduced in its true sense. In its twelve chapters this report analysed and provided a clear overview of the conditions in the world (socio-economic development and order, environmental degradation, population growth, poverty, politics, wars, etc.) and elaborated the concept of sustainable development. As a new approach, this concept should be able to respond to future challenges, such as achieving balance between socio-economic development and the environment, reducing pollution and environmental degradation, exploiting natural resources, reducing harmful gas emissions and climate impacts, reducing poverty and hunger, achieving world peace and other serious challenges and threats faced by humanity. The concept of sustainable development is defined as "development that meets the needs of the present without compromising the ability of future generations to meet their own needs", which contains the core of the concept and soon became a generally accepted and probably the most cited definition in the literature, no matter where the context of sustainable development is being discussed.

The fundamental objective of the concept outlined in the document is to provide basic human needs to all people (home, food, water, clothing, etc.), with a tendency to improve living standards, as well to achieve the aspiration of a better life. An imperative of the Brundtland report is: rational and controlled use of resources focused on renewable and long-term usage, protection and conservation of nature, raising ecological awareness, stricter national regulation and international co-operation, stopping population growth, using industry and technology in line with environmental requirements, developing technological innovations in order to reduce impact on environmental. Thus, according to the report, the underlying principles of the concept of sustainable development are assurance of the human needs, while respecting certain environmental constraints. The Brundtland report marked the beginning of a new global socio-economic policy in which the concept of sustainable development has become a key element in environmental management and other areas of human activities.

This event was followed by the third, so-called After Brundtland period, which lasts until today and included several significant events. Marking the twentieth anniversary of the conference in Stockholm, UN conference on environment and development called the Earth Summit or the Rio Conference was held in Rio de Janeiro in 1992. The conference saw the participation of numerous governmental and non-governmental organizations from 178 countries. Its focus was to define a global framework for solving issues of environmental degradation through the concept of sustainable development, considering that in the 20-year period the integration of environmental concerns and economic decision-making was ignored and the state of the environment was worse. More than 10,000 international journalists transmitted the conference to millions of people around the world, witnessing the importance of the conference. The preparation of the conference began in 1989 and as a result the following documents were adopted: 1) Rio Declaration on Environment and Development, 2) Agenda 21, 3) Non-legally binding authoritative statement of principles for a global consensus on the management, conservation and sustainable development of all types of forests, 4) Climate Change Convention and 5) Convention on Biological Diversity. The first two documents are key for the concept of sustainable development.

The Rio Declaration on Environment and Development contains 27 principles of sustainable development on the rights and responsibilities of the United Nations. These principles also form the basis for future policy and decision making and balance between socio-economic development and the environment. The Declaration gives people the right for development but also the obligation for preserving the environment, and since the environment is a public and common good, it also highlights the need for cooperation and understanding between the public and private sectors and civil society. Among the principles, it is emphasized how humans are in the centre of concern for sustainable development and should not delay measures to prevent environmental degradation. At the same time, it is emphasized that each country has the sovereign right to exploit its own resources, if this does not endanger the environment of other countries, thereby polluters should bear the costs of pollution. Eradication of poverty, reduction of inequalities and assuring basic living standards and peace in the world are essential for sustainable development, therefore developed countries have the responsibility to ensure sustainable development, particularly for technology and financial resources.

Agenda 21 is a global programme with objectives of sustainable development and action plans. The document comprehensively provides guidelines for socio-economic development in line with the environmental conservation. The document highlights the need for international cooperation and consensus between development and environmental protection, whereby governments play an important role in the adoption and implementation of policies, plans and programmes, although the participation of all other stakeholders is also necessary. Further on, developed countries play a key role, particularly in providing financial funds to developing countries. As a priority goal, the document emphasizes the suppression of poverty, especially in poor countries where it is also necessary to preserve and protect natural resources. At the same time, in these countries there is a need for improvement of the protection of human health and gender equality. It is also necessary to change patterns of behaviour in production and consumption in order to rationally exploit natural resources and fossil fuels which would result in reduced negative impact on the environment. Finally, Agenda 21 highlights the importance of educational programmes focused on raising awareness and promotion of the sustainable development which are necessary for its implementation.

From these fundamental activities and documents the three key elements of the concept were identified: 1) the concept of development (socio-economic development in line with ecological constraints), 2) the concept of needs (redistribution of resources to ensure the quality of life for all) and 3) the concept of future generations (the possibility of a long-term usage of resources to ensure the necessary quality of life for future generations). At the same time, concept of sustainable development outlined core principles, namely: ensuring needs and care for the community of present and future generations, continuously improving the overall quality of life and equality, protecting and preserving the environment, biodiversity and ecosystems, protecting and preserving the natural resources, with the rational use of renewable resources and reduced depletion of non-renewable resources, changing production and consumption respecting the ecological constraints, using renewable energy and innovative technologies to reduce the negative impact on the environment, strengthening international cooperation at the national, regional and local level, creating an institutional framework with a strong network of stakeholders interested in implementing the concept of sustainable development, etc. Here it could be mentioned how the three key elements of the concept were also described by the Maslowian portfolio theory (MaPT) and the hierarchy of needs.

SUSTAINABLE DEVELOPMENT GOALS

The Sustainable Development Goals (SDGs) are a collection of 17 global goals set by the United Nations General Assembly in 2015 for the year 2030. The SDGs are part of Resolution 70/1 of the United Nations General Assembly, the 2030 Agenda.

The Sustainable Development Goals are:

1. No Poverty,

2. Zero Hunger,

3. Good Health and Well-being,

4. Quality Education,

5. Gender Equality,

6. Clean Water and Sanitation,

7. Affordable and Clean Energy,

8. Decent Work and Economic Growth,

9. Industry, Innovation, and Infrastructure,

10. Reducing Inequality,

11. Sustainable Cities and Communities,

12. Responsible Consumption and Production,

13. Climate Action,

14. Life Below Water,

15. Life on Land,

16. Peace, Justice, and Strong Institutions,

17. Partnerships for the Goals.

The goals are broad based and interdependent. The 17 sustainable development goals each have a list of targets that are measured with indicators.

Key to making the SDGs successful is to make the data on the 17 goals available and understandable. Various tools exist to track and visualize progress towards the goals.

UN SDG consultations in Mariupol, Ukraine.

In 1972, governments met in Stockholm, Sweden for the United Nations Conference on the Human Environment, to consider the rights of the family to a healthy and productive environment. In 1983, the United Nations created the World Commission on Environment and Development (later known as the Brundtland Commission), which defined sustainable development as "meeting the needs of the present without compromising the ability of future generations to meet their own needs". In 1992, the first United Nations Conference on Environment and Development (UNCED) or Earth Summit was held in Rio de Janeiro, where the first agenda for Environment and Development, also known as Agenda 21, was developed and adopted.

In 2012, the United Nations Conference on Sustainable Development (UNCSD), also known as Rio+20, was held as a 20-year follow up to UNCED. Colombia proposed the idea of the SDGs at a preparation event for Rio+20 held in Indonesia in July 2011. In September 2011, this idea was picked up by the United Nations Department of Public Information 64th NGO Conference in Bonn, Germany. The outcome document proposed 17 sustainable development goals and associated targets. In the run-up to Rio+20 there was much discussion about the idea of the SDGs. At the Rio+20 Conference, a resolution known as "The Future We Want" was reached by member states. Among the key themes agreed on were poverty eradication, energy, water and sanitation, health, and human settlement.

The Rio+20 outcome document mentioned that "at the outset, the OWG (Open Working Group) will decide on its methods of work, including developing modalities to ensure the full involvement of relevant stakeholders and expertise from civil society, Indigenous Peoples, the scientific community and the United Nations system in its work, in order to provide a diversity of perspectives and experience".

The Post-2015 Development Agenda was a process from 2012 to 2015 led by the United Nations to define the future global development framework that would succeed the Millennium Development Goals. The SDGs were developed to succeed the Millennium Development Goals (MDGs) which ended in 2015. The gaps and shortcomings of MDG Goal 8 (To develop a global partnership for development) led to identifying a problematic "donor-recipient" relationship. Instead, the new SDGs favor collective action by all countries.

The UN-led process involved its 193 Member States and global civil society. The resolution is a broad intergovernmental agreement that acts as the Post-2015 Development Agenda. The SDGs build on the principles agreed upon in Resolution A/RES/66/288, entitled "The Future We Want". This was a non-binding document released as a result of Rio+20 Conference held in 2012.

Goal 1: No Poverty

"End poverty in all its forms everywhere."

Extreme poverty has been cut by more than half since 1990. Still, around 1 in 10 people live on less than the target figure of international-$1.90 per day. A very low poverty threshold is justified by highlighting the need of those people who are worst off. SDG 1 is to end extreme poverty globally by 2030.

That target may not be adequate for human subsistence and basic needs, however. It is for this reason that changes relative to higher poverty lines are also commonly tracked. Poverty is more than the lack of income or resources: People live in poverty if they lack basic services such as healthcare, security, and education. They also experience hunger, social discrimination, and exclusion from decision-making processes. One possible alternative metric is the Multidimensional Poverty Index.

Children make up the majority – more than half – of those living in extreme poverty. In 2013, an estimated 385 million children lived on less than US$1.90 per day. Still, these figures are unreliable due to huge gaps in data on the status of children worldwide. On average, 97 percent of countries have insufficient data to determine the state of impoverished children and make projections towards SDG Goal 1, and 63 percent of countries have no data on child poverty at all.

Women face potentially life-threatening risks from early pregnancy and frequent pregnancies. This can result in lost hope for an education and for a better income. Poverty affects age groups differently, with the most devastating effects experienced by children. It affects their education, health, nutrition, and security, impacting emotional and spiritual development.

Achieving Goal 1 is hampered by lack of economic growth in the poorest countries of the world, growing inequality, increasingly fragile statehood, and the impacts of climate change.

Goal 2: Zero Hunger

"End hunger, achieve food security and improved nutrition, and promote sustainable agriculture".

Goal 2 states that by 2030 we should end hunger and all forms of malnutrition. This would be accomplished by doubling agricultural productivity and incomes of small-scale food producers (especially women and indigenous peoples), by ensuring sustainable food production systems, and by progressively improving land and soil quality. Agriculture is the single largest employer in the world, providing livelihoods for 40% of the global population. It is the largest source of income for poor rural households. Women make up about 43% of the agricultural labor force in developing countries, and over 50% in parts of Asia and Africa. However, women own only 20% of the land.

Other targets deal with maintaining genetic diversity of seeds, increasing access to land, preventing trade restriction and distortions in world agricultural markets to limit extreme food price volatility, eliminating waste with help from the International Food Waste Coalition, and ending malnutrition and undernutrition of children.

Globally, 1 in 9 people are undernourished, the vast majority of whom live in developing countries. Undernutrition causes wasting or severe wasting of 52 million children worldwide, and contributes to nearly half (45%) of deaths in children under five – 3.1 million children per year. Chronic malnutrition, which affects an estimated 155 million children worldwide, also stunts children's brain and physical development and puts them at further risk of death, disease, and lack of success as adults. As of 2017, only 26 of 202 UN member countries are on track to meet the SDG target to eliminate undernourishment and malnourishment, while 20 percent have made no progress at all and nearly 70 percent have no or insufficient data to determine their progress.

A report by the International Food Policy Research Institute (IFPRI) of 2013 stated that the emphasis of the SDGs should not be on ending poverty by 2030, but on eliminating hunger and under-nutrition by 2025. The assertion is based on an analysis of experiences in China, Vietnam, Brazil, and Thailand. Three pathways to achieve this were identified: 1) agriculture-led; 2) social protection- and nutrition- intervention-led; or 3) a combination of both of these approaches.

Goal 3: Good Health and Well-being for People

"Ensure healthy lives and promote well-being for all at all ages."

Significant strides have been made in increasing life expectancy and reducing some of the common killers associated with child and maternal mortality. Between 2000 and 2016, the worldwide under-five mortality rate decreased by 47 percent (from 78 deaths per 1,000 live births to 41 deaths per 1,000 live births). Still, the number of children dying under age five is extremely high: 5.6 million in 2016 alone. Newborns account for a growing number of these deaths, and poorer children are at the greatest risk of under-5 mortality due to a number of factors. SDG Goal 3 aims to reduce under-five mortality to at least as low as 25 per 1,000 live births. But if current trends continue, more than 60 countries will miss the SDG neonatal mortality target for 2030. About half of these countries would not reach the target even by 2050.

Goal 3 also aims to reduce maternal mortality to less than 70 deaths per 100,000 live births. Though the maternal mortality ratio declined by 37 percent between 2000 and 2015, there were approximately 303,000 maternal deaths worldwide in 2015, most from preventable causes. In 2015, maternal health conditions were also the leading cause of death among girls aged 15-19. Data for girls of greatest concern – those aged between 10-14 - is currently unavailable. Key strategies for meeting SDG Goal 3 will be to reduce adolescent pregnancy (which is strongly linked to gender equality), provide better data for all women and girls, and achieve universal coverage of skilled birth attendants.

Similarly, progress has been made on increasing access to clean water and sanitation and on reducing malaria, tuberculosis, polio, and the spread of HIV/AIDS. From 2000-2016, new HIV infections declined by 66 percent for children under 15 and by 45 percent among adolescents aged 15-19. However, current trends mean that 1 out of 4 countries still won't meet the SDG target to end AIDS among children under 5, and 3 out of 4 will not meet the target to end AIDS among adolescents. Additionally, only half of women in developing countries have received the health care they need, and the need for family planning is increasing exponentially as the population grows. While needs are being addressed gradually, more than 225 million women have an unmet need for contraception.

Goal 3 aims to achieve universal health coverage, including access to essential medicines and vaccines. It proposes to end the preventable death of newborns and children under 5 and to end epidemics such as AIDS, tuberculosis, malaria, and water-borne diseases, for example. 2016 rates for the third dose of the pertussis vaccine (DTP3) and the first dose of the measles vaccine (MCV1) reached 86 percent and 85 percent, respectively. Yet about 20 million children did not receive DTP3 and about 21 million did not receive MCV1. Around 2 in 5 countries will need to accelerate progress in order to reach SDG targets for immunization.

Attention to health and well-being also includes targets related to the prevention and treatment of substance abuse, deaths and injuries from traffic accidents and from hazardous chemicals and air, water and soil pollution and contamination.

Goal 4: Quality Education

"Ensure inclusive and equitable quality education and promote lifelong learning opportunities for all."

Major progress has been made in access to education, specifically at the primary school level, for both boys and girls. The number of out-of-school children has almost halved from 112 million in 1997 to 60 million in 2014. Still, at least 22 million children in 43 countries will miss out on pre-primary education unless the rate of progress doubles.

Access does not always mean quality of education or completion of primary school. 103 million youth worldwide still lack basic literacy skills, and more than 60 percent of those are women. In one out of four countries, more than half of children failed to meet minimum math proficiency standards at the end of primary school, and at the lower secondary level, the rate was 1 in 3 countries. Target 1 of Goal 4 is to ensure that, by 2030, all girls and boys complete free, equitable, and quality primary and secondary education.

Additionally, progress is difficult to track: 75 percent of countries have no or insufficient data to track progress towards SDG Goal 4 targets for learning outcomes (target 1), early childhood education (target 2), and effective learning environments. Data on learning outcomes and pre-primary school are particularly scarce; 70 percent and 40 percent of countries lack adequate data for these targets, respectively. This makes it hard to analyze and identify the children at greatest risk of being left behind.

Goal 5: Gender Equality

"Achieve gender equality and empower all women and girls."

According to the UN, "gender equality is not only a fundamental human right, but a necessary foundation for a peaceful, prosperous and sustainable world." Providing women and girls with equal access to education, health care, decent work, and representation in political and economic decision-making processes will nurture sustainable economies and benefit societies and humanity at large. A record 143 countries guaranteed equality between men and women in their constitutions as of 2014. However, another 52 had not taken this step. In many nations, gender discrimination is still woven into the fabric of legal systems and social norms. Even though SDG5 is a standalone goal, other SDGs can only be achieved if the needs of women receive the same attention as the needs of men. Issues unique to women and girls include traditional practices against all women and girls in the public and private spheres, such as female genital mutilation.

Child marriage has declined over the past decades, yet there is no region that is currently on track to eliminate the practice and reach SDG targets by 2030. If current trends continue, between 2017 and 2030, 150 million girls will be married before they turn 18. Though child marriages are four times higher among the poorest than the wealthiest in the world, most countries need to accelerate progress among both groups in order to reach the SDG Goal 5 target to eliminate child marriage by 2030.

Achieving gender equality will require enforceable legislation that promotes empowerment of all women and girls and requires secondary education for all girls. The targets call for an end to gender discrimination and for empowering women and girls through technology Some have advocated for "listening to girls". The assertion is that the SDGs can deliver transformative change for girls only if girls are consulted. Their priorities and needs must be taken into account. Girls should be viewed not as beneficiaries of change, but as agents of change. Engaging women and girls in the implementation of the SDGs is crucial.

The World Pensions Council (WPC) has insisted on the transformational role gender-diverse that boards can play in that regard, predicting that 2018 could be a pivotal year, as "more than ever before, many UK and European Union pension trustees speak enthusiastically about flexing their fiduciary muscles for the UN's Sustainable Development Goals, including SDG5, and to achieve gender equality and empower all women and girls."

Goal 6: Clean Water and Sanitation

Example of sanitation for all: School toilet (IPH school and college, Mohakhali, Dhaka, Bangladesh).

Unimproved sanitation example: pit latrine without slab in Lusaka, Zambia.

"Ensure availability and sustainable management of water and sanitation for all."

The Sustainable Development Goal Number 6 (SDG6) has eight targets and 11 indicators that will be used to monitor progress toward the targets. Most are to be achieved by the year 2030. One is targeted for 2020.

The first three targets relate to drinking water supply and sanitation. Worldwide, 6 out of 10 people lack safely managed sanitation services, and 3 out of 10 lack safely managed water services. Safe drinking water and hygienic toilets protect people from disease and enable societies to be more productive economically. Attending school and work without disruption is critical to successful education and successful employment. Therefore, toilets in schools and work places are specifically mentioned as a target to measure. "Equitable sanitation" calls for addressing the specific needs of women and girls and those in vulnerable situations, such as the elderly or people with disabilities.

Water sources are better preserved if open defecation is ended and sustainable sanitation systems are implemented.

Ending open defecation will require provision of toilets and sanitation for 2.6 billion people as well as behavior change of the users. This will require cooperation between governments, civil society, and the private sector.

The main indicator for the sanitation target is the "Proportion of population using safely managed sanitation services, including a hand-washing facility with soap and water". However, as of 2017, two-thirds of countries lacked baseline estimates for SDG indicators on hand washing, safely managed drinking water, and sanitation services. From those that were available, the Joint Monitoring Programme (JMP) found that 4.5 billion people currently do not have safely managed sanitation. If we are to meet SDG targets for sanitation by 2030, nearly one-third of countries will need to accelerate progress to end open defecation including Brazil, China, Ethiopia, India, Indonesia, Nigeria, and Pakistan.

The Sustainable Sanitation Alliance (SuSanA) has made it its mission to achieve SDG6. SuSanA's position is that the SDGs are highly interdependent. Therefore, the provision of clean water and sanitation for all is a precursor to achieving many of the other SDGs.

Goal 7: Affordable and Clean Energy

"Ensure access to affordable, reliable, sustainable and modern energy for all."

Targets for 2030 include access to affordable and reliable energy while increasing the share of renewable energy in the global energy mix. This would involve improving energy efficiency and enhancing international cooperation to facilitate more open access to clean energy technology and more investment in clean energy infrastructure. Plans call for particular attention to infrastructure support for the least developed countries, small islands and land-locked developing countries.

As of 2017, only 57 percent of the global population relies primarily on clean fuels and technology, falling short of the 95 percent target.

Goal 8: Decent Work and Economic Growth

"Promote sustained, inclusive and sustainable economic growth, full and productive employment and decent work for all."

World Pensions Council (WPC) development economists have argued that the twin considerations of long-term economic growth and infrastructure investment weren't prioritized enough. The fact they were designated as the number 8 and number 9 objective respectively was considered a rather "mediocre ranking which defies common sense".

For the least developed countries, the economic target is to attain at least a 7 percent annual growth in gross domestic product (GDP). Achieving higher productivity will require diversification and upgraded technology along with innovation, entrepreneurship, and the growth of small- and medium-sized enterprises (SMEs). Some targets are for 2030; others are for 2020.

The target for 2020 is to reduce youth unemployment and operationalize a global strategy for youth employment.

By 2030, the target is to establish policies for sustainable tourism that will create jobs. Strengthening domestic financial institutions and increasing Aid for Trade support for developing countries is considered essential to economic development. The Enhanced Integrated Framework for Trade-Related Technical Assistance to Least Developed Countries is mentioned as a method for achieving sustainable economic development.

Goal 9: Industry, Innovation and Infrastructure

"Build resilient infrastructure, promote inclusive and sustainable industrialization, and foster innovation".

Manufacturing is a major source of employment. In 2016, the least developed countries had less "manufacturing value added per capita". The figure for Europe and North America amounted to US$4,621, compared to about $100 in the least developed countries. The manufacturing of high products contributes 80 percent to total manufacturing output in industrialized economies but barely 10 percent in the least developed countries.

Mobile-cellular signal coverage has improved a great deal. In previously "unconnected" areas of the globe, 85 percent of people now live in covered areas. Planet-wide, 95 percent of the population is covered.

Goal 10: Reducing Inequalities

"Reduce income inequality within and among countries."

Target 10.1 is to "sustain income growth of the bottom 40 per cent of the population at a rate higher than the national average". This goal, known as 'shared prosperity', is complementing SDG 1, the eradication of extreme poverty, and it is relevant for all countries in the world.

Target 10.3 is to reduce the transaction costs for migrant remittances to below 3 percent. The target of 3 percent was established as the cost that international migrant workers would pay to send money home (known as remittances). However, post offices and money transfer companies currently charge 6 percent of the amount remitted. Worse, commercial banks charge 11 percent. Prepaid cards and mobile money companies charge 2 to 4 percent, but those services were not widely available as of 2017 in typical "remittance corridors."

Goal 11: Sustainable Cities and Communities

"Make cities and human settlements inclusive, safe, resilient, and sustainable."

The target for 2030 is to ensure access to safe and affordable housing. The indicator named to measure progress toward this target is the proportion of urban population living in slums or informal settlements. Between 2000 and 2014, the proportion fell from 39 percent to 30 percent. However, the absolute number of people living in slums went from 792 million in 2000 to an estimated 880 million in 2014. Movement from rural to urban areas has accelerated as the population has grown and better housing alternatives are available.

Goal 12: Responsible Consumption and Production

"Ensure sustainable consumption and production patterns."

The targets of Goal 12 include using eco-friendly production methods and reducing the amount of waste. By 2030, national recycling rates should increase, as measured in tons of material recycled. Further, companies should adopt sustainable practices and publish sustainability reports.

Target 12.1 calls for the implementation of the 10-year Framework of Programmes on Sustainable Consumption and Production. This framework, adopted by member states at the United Nations Conference on Sustainable Development, is a global commitment to accelerate the shift to sustainable consumption and production in developed and developing countries. In order to generate the collective impact necessary for such a shift, programs such as the One Planet Network have formed different implementation methods to help achieve Goal 12.

Goal 13: Climate Action

"Take urgent action to combat climate change and its impacts by regulating emissions and promoting developments in renewable energy."

The UN discussions and negotiations identified the links between the post-2015 SDG process and the Financing for Development process that concluded in Addis Ababa in July 2015 and the COP 21 Climate Change conference in Paris in December 2015.

In May 2015, a report concluded that only a very ambitious climate deal in Paris in 2015 could enable countries to reach the sustainable development goals and targets. The report also states that tackling climate change will only be possible if the SDGs are met. Further, economic development and climate change are inextricably linked, particularly around poverty, gender equality, and energy. The UN encourages the public sector to take initiative in this effort to minimize negative impacts on the environment.

This renewed emphasis on climate change mitigation was made possible by the partial Sino-American convergence that developed in 2015-2016, notably at the UN COP21 summit (Paris) and ensuing G20 conference (Hangzhou).

As one of the regions most vulnerable to the effects of climate change, the Asia-Pacific region needs more public-private partnerships (PPPs) to successfully implement its sustainable development initiatives.

In 2018, the International Panel of Climate Change (IPCC), the United Nations body for assessing the science related to climate change, published a special report *"Global Warming of 1.5 °C"*. It outlined the impacts of a 1.5 °C global temperature rise above pre-industrial levels and related global greenhouse gas emission pathways, and highlighted the possibility of avoiding a number of such impacts by limiting global warming to 1.5 °C compared to 2°C, or more. The report mentioned that this would require global net human-caused emissions of carbon dioxide (CO_2) to fall by about 45% from 2010 levels by 2030, reaching "net zero" around 2050, through "rapid and far-reaching" transitions in land, energy, industry, buildings, transport, and cities. This special report was subsequently discussed at COP 24. Despite being requested by countries at the COP 21, the report was not accepted by four countries – the US, Saudi Arabia, Russia and Kuwait, which only wanted to "note" it, thereby postponing the resolution to the next SBSTA session in 2019.

Goal 14: Life Below Water

"Conserve and sustainably use the oceans, seas and marine resources for sustainable development."

Sustainable Development Goal 14 aims "to conserve and sustainably use the oceans, seas and marine resources for sustainable development." Effective strategies to mitigate adverse effects of increased ocean acidification are needed to advance the sustainable use of oceans. As areas of protected marine biodiversity expand, there has been an increase in ocean science funding, essential for preserving marine resources. The deterioration of coastal waters has become a global occurrence, due to pollution and coastal eutrophication (overflow of nutrients in water), where similar contributing factors to climate change can affect oceans and negatively impact marine biodiversity. "Without concerted efforts, coastal eutrophication is expected to increase in 20 per cent of large marine ecosystems by 2050."

The Preparatory Meeting to the UN Ocean Conference convened in New York, US, in February 2017, to discuss the implementation of Sustainable Development Goal 14. International law, as reflected in the UN Convention on the Law of the Sea (UNCLOS), stressed the need to include governance instruments to consider "anthropogenic activities taking place outside of the ocean". Concerns regarding ocean health in destructive fishing practices and marine pollution were discussed, in looking at the role of local communities of small island developing States (SIDS) and least developed countries (LDCs) to not forget that oceans are a large part of their economies. The targets include preventing and reducing marine pollution and acidification, protecting marine and coastal ecosystems, and regulating fishing. The targets also call for an increase in scientific knowledge of the oceans.

Although many participating United Nations legislative bodies comes together to discuss the issues around marine environments and SDG 14, such as at the United Nations Ocean Conference, it is important to consider how SDG 14 is implemented across different Multilateral Environmental Agreements, respectively. As climate, biodiversity and land degradation are major parts of the issues surrounding the deterioration of marine environments and oceans, it is important to know how each Rio Convention implements this SDG.

Oceans cover 71 percent of the Earth's surface. They are essential for making the planet livable. Rainwater, drinking water and climate are all regulated by ocean temperatures and currents. Over 3 billion people depend on marine life for their livelihood. Oceans absorb 30 percent of all carbon dioxide produced by humans. The oceans contain more than 200,000 identified species, and there might be thousands of species that are yet to be discovered. Oceans are the world's largest sources of protein. However, there has been a 26 percent increase in acidification since the industrial revolution. A full 30 percent of marine habitats have been destroyed, and 30 percent of the world's fish stocks are over-exploited. Marine pollution has reached shocking levels; each minute, 15 tons of plastic are released into the oceans. 20 percent of all coral reefs have been destroyed irreversibly, and another 24 percent are in immediate risk of collapse.

Approximately, 1 million sea birds, 100 000 marine mammals, and an unknown number of fish are harmed or die annually due to marine pollution caused by humans. It has been found that 95 percent of fulmars in Norway have plastic parts in their guts. Microplastics are another form of marine pollution.

Individuals can help the oceans by reducing their energy consumption and their use of plastics. Nations can also take action. In Norway, for instance, citizens, working through a web page called finn.no, can earn money for picking up plastic on the beach. Several countries, including Kenya, have banned the use of plastic bags for retail purchases. Improving the oceans contributes to poverty reduction, as it gives low-income families a source of income and healthy food. Keeping beaches and ocean water clean in less developed countries can attract tourism, as stated in Goal 8, and reduce poverty by providing more employment.

Characterized by extinctions, invasions, hybridizations and reductions in the abundance of species, marine biodiversity is currently in global decline. "Over the past decades, there has been an exponential increase in human activates in and near oceans, resulting in negative consequences to our marine environment." Made evident by the degradation of habitats and changes in ecosystem processes, the declining health of the oceans has a negative effect on people, their livelihoods and entire economies, with local communities that rely on ocean resources being the most affected. Poor decisions in resource management can compromise conservation, local livelihood, and resource sustainability goals. "The sustainable management of our oceans relies on the ability to influence and guide human use of the marine environment." As conservation of marine resources is critical to the well-being of local fishing communities and their livelihoods, related management actions may lead to changes in human behaviour to support conservation programs to achieve their goals. Ultimately, governments and international agencies act as gatekeepers, interfering with needed stakeholder participation in decision making. The way to best safeguard life in oceans is to implement effective management strategies around marine environments.

Climate action is used as a way of protecting the world's oceans. Oceans cover three quarters of the Earth's surface and impact global climate systems through functions of carbon dioxide absorption from the atmosphere and oxygen generation. The increase in levels of greenhouse gases leading to changes in climate negatively affects the world's oceans and marine coastal communities. The resulting impacts of rising sea levels by 20 centimeters since the start of the 20th century and the increase of ocean acidity by 30% since the Industrial Revolution has contributed to the melting of ice sheets through the thermal expansion of sea water.

Sustainable Development Goal 14 has been incorporated into the Convention on Biological Diversity (CBD), the United Nations Framework Convention on Climate Change (UNFCCC), and the United Nations on Combat Desertification (UNCCD).

Goal 15: Life on Land

"Protect, restore and promote sustainable use of terrestrial ecosystems, sustainably manage forests, combat desertification, and halt and reverse land degradation and halt biodiversity loss."

This goal articulates targets for preserving biodiversity of forest, desert, and mountain eco-systems, as a percentage of total land mass. Achieving a "land degradation-neutral world" can be reached by restoring degraded forests and land lost to drought and flood. Goal 15 calls for more attention to preventing invasion of introduced species and more protection of endangered species. Forests have a prominent role to play in the success of Agenda 2030, notably in terms of ecosystem services, livelihoods, and the green economy; but this will require clear priorities to address key tradeoffs and mobilize synergies with other SDGs.

The Mountain Green Cover Index monitors progress toward target 15.4, which focuses on preserving mountain ecosystems. The index is named as the indicator for target 15.4. Similarly, the Red Index will fill the monitoring function for biodiversity goals by documenting the trajectory of endangered species. Animal extinction is a growing problem.

Goal 16: Peace, Justice and Strong Institutions

"Promote peaceful and inclusive societies for sustainable development, provide access to justice for all and build effective, accountable and inclusive institutions at all levels."

Reducing violent crime, sex trafficking, forced labor, and child abuse are clear global goals. The International Community values peace and justice and calls for stronger judicial systems that will enforce laws and work toward a more peaceful and just society. By 2017, the UN could report progress on detecting victims of trafficking. More women and girls than men and boys were victimized, yet the share of women and girls has slowly declined. In 2004, 84 percent of victims were females and by 2014 that number had dropped to 71 percent. Sexual exploitation numbers have declined, but forced labor has increased.

One target is to see the end to sex trafficking, forced labor, and all forms of violence against and torture of children. However, reliance on the indicator of "crimes reported" makes monitoring and achieving this goal challenging. For instance, 84 percent of countries have no or insufficient data on violent punishment of children. Of the data available, it is clear that violence against children by their caregivers remains pervasive: Nearly 8 in 10 children aged 1 to 14 are subjected to violent discipline on a regular basis (regardless of income), and no country is on track to eliminate violent discipline by 2030.

SDG 16 also targets universal legal identity and birth registration, ensuring the right to a name and nationality, civil rights, recognition before the law, and access to justice and social services. With more than a quarter of children under 5 unregistered worldwide as of 2015, about 1 in 5 countries will need to accelerate progress to achieve universal birth registration by 2030.

Goal 17: Partnerships for the Goals

"Strengthen the means of implementation and revitalize the global partnership for sustainable development."

Increasing international cooperation is seen as vital to achieving each of the 16 previous goals. Goal 17 is included to assure that countries and organizations cooperate instead of compete. Developing multi-stakeholder partnerships to share knowledge, expertise, technology, and financial support is seen as critical to overall success of the SDGs.

SUSTAINABLE DEVELOPMENT INDICATORS

The need for reliable and pertinent indicators to guide the sustainable development process was recognised early, at the time of the Rio Conference. It was reaffirmed in many sections of Agenda 21 the programme document which was agreed at the summit, which deals with information required for decision-making.

> "Commonly used indicators such as the gross national product (GNP) and measurements of individual resource or pollution flows do not provide adequate indications of sustainability. Methods for assessing interactions between different sectoral environmental, demographic, social and developmental parameters are not sufficiently developed or applied. Indicators of sustainable development need to be developed to provide solid bases for decision-making at all levels and to contribute to a self-regulating sustainability of integrated environment and development systems."

Therefore,

> "Countries and international organizations should review and strengthen information systems and services in sectors related to sustainable development, at the local, provincial, national and international levels. Special emphasis should be placed on the transformation of existing information into forms more useful for decision-making and on targeting information at different user groups. Mechanisms should be strengthened or established for transforming scientific and socio-economic assessments into information suitable for both planning and public information. Electronic and non-electronic formats should be used."

In the opinion of the authors of Agenda 21, current indicators (including GDP) are incapable of evaluating the "sustainability of systems". Furthermore, existing information cannot be used in this format for decision-making and must be converted and then redirected at the various user groups. Several questions are left unanswered, to which the authors of Agenda 21 would have us reply. Who are these groups of users? Into what forms, more appropriate for decision-making, should the information be converted? How should it be converted for use in decision-making? What sectors are involved in sustainable development? In the following topic, we will be suggesting a few pointers to respond to these questions and some indications on the construction of appropriate information systems for sustainable development, i.e. adequate, pertinent and acceptable to all development actors. In the space available, it will not be possible to provide sufficiently detailed and qualified considerations of these issues, so that certain simplifications will have to be used, at the risk of painting with a broad brush at times. For example, the subject of the various user groups will be dealt with in a voluntarily reductive fashion, based on the following question "Indicators for whom: governments or citizens?" The question on the more or less usable forms will be limited to asking "scoreboard or synthetic indices?". And the question of sectors involved in sustainable development will be reduced to a comparison between four major approaches to the actual object of sustainable development.

Indicators: Scoreboard or Synthetic Index

The concept of indicators was originally used in a purely scientific context: sociological research. It designated the translation of theoretical (abstract) concepts into observable variables so that the scientific hypotheses involving these concepts could be submitted to empirical verification. We come across the word in a seminal text by Lazarsfeld on the operationalisation of sociological theories where the various stages in the translation of concepts into indices were clearly identified and analysed for the first time.

An indicator is therefore an observable variable used to report a non-observable reality. As regards the word 'index', it designates a synthetic indicator constructed by aggregating other so-called 'basic" indicators. Most of the indicators used in public policy-making are in fact indices: this is true for GDP, the index of consumer prices, stock exchange indices such as the Dow-Jones and the Human Development Index (HDI) of the United Nationals Development Programme (UNDP).

Shortly after Lazarsfeld's work was published, the word 'indicator", to which the 'social' was added as a qualifier, became popular in the public domain, or at least in the domain of public policy. A "social indicator movement" emerged in the United States, then in Europe, following the publication by Bauer, Biderman and Gross of a report called "Social Indicators". Whereas for Lazarsfeld and later, the scientific community, the role of indicators was purely methodological, it became normative and axiological with the movement for social indicators. The reference to norms and values is given at the outset in the definition Bauer gives for social indicators: "statistics, statistical series, and all other forms of evidence that enable us to assess where we stand and are going with respect to our values and goals."

While the term "indicator" was new, the reality described was much older, not to say immemorial. The same term in fact covered two traditions, one, age-old and the other going back to the industrial revolution. The first is the concept of statistics in the original meaning of the word, i.e. the methodical study of social facts by numerical processes (classifications, counting, quantified inventories and censuses) for the purpose of information and assisting governments. The other more recent source is to be found in the numerous movements for social reform and hygiene at the time of the industrial revolution. At the start of the 19th century, philanthropists (often physicians or clergymen) were using statistical data on housing, living and working conditions, income, alcoholism, prisons, etc. with the aim of reforming society and improving the lot of the underprivileged. In the United States, the first known use of social indicators for the purpose of social reform goes back to around 1810, with the production of statistical data for five consecutive years on the number of inmates awaiting trial in Philadelphia prisons. Other surveys are well-known, such as those on poverty by Villermé in France, Ducpétiaux in Belgium and Booth in the U.K.

After the decline of the social indicators movement of the sixties, the concept of social indicator suffered a lapse of several decades before re-emerging quite recently, first with reference to the measurement of human welfare and development and later with reference to the notion of sustainability and sustainable development. Observers, among them Gadrey and Jany-Catrice, Perret and Sharpe were numerous in remarking on the recent proliferation of attempts—if not at replacing GDP—at least supplementing it with a more adequate synthetic measurement of well-being.

Among these attempts, only one achieved a real measure of success: this was the UNDP Human Development Index. All the others—be it the ISEW (Index of Sustainable Economic Welfare) created by Daly and Cobb, the GPI the MDP, the Index of Economic Well-being created by Sharpe and Osberg, the HWI, etc.—failed to gain much favour or sufficient legitimacy to become institutionalised.

The exception represented by the Human Development Index is rather enlightening: without the backing of the Nobel Prize for Economic Science laureate Amartya Sen, it probably would also have failed to pass muster. On closer examination, it is not so much indicators that come up against a degree of opposition (in particular from the scientific community) but rather indices or synthetic indicators. There is no opposition, quite the contrary, to the proliferation of scoreboards of every variety, i.e. batteries of indicators, be it in the environmental or the "social" sectors. However, the construction of indices, in particular the Human Development Index, sets off reactions such as the one by Baneth, for example, who goes so far as to say: "It was a vain, pretentious and slightly ridiculous endeavour to try to sum up human development in all its complexity and multiple dimensions with a single figure.

And yet the only difference between a management chart and a synthetic index lies in the ultimate phase of the construction and measuring process of the indicators: that is the production, using basic indicators, of a single synthetic value for the purpose of condensing the information contained in the management chart. In other words, a synthetic index is no more or less than a scoreboard to which is added an extra indicator made up of the aggregation of the data contained in it. But it would seem that for some people, this ultimate phase is all the difference between a rigorously serious and scientific effort and a subjective, ideological and fanciful exercise.

The Construction of Indicators

The Successive Phases

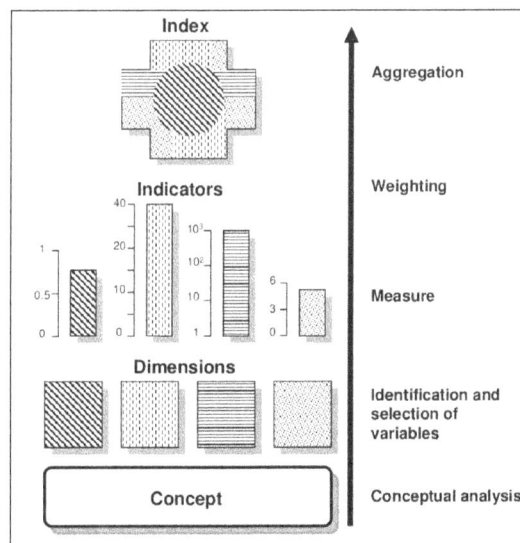

From concept to indices.

Figure shows the successive phases of the construction of indicators identified by Lazarsfeld.

From Concept to Dimensions

The first phase consists in identifying the various dimensions constituting the concept, given that these are always multidimensional. The concept of poverty, for example, covers a material dimension, but also a social one (exclusion, marginalisation) and also a cultural dimension (level of education, means of expression). The material dimension is itself multi-faceted; it includes financial components (income, level of indebtedness, other financial burdens) and non-financial ones (health, housing, rights). Each of these material dimensions is itself more or less composite. Income, for instance, may or may not be monetary. A further point is that the regular or precarious nature of income matters more sometimes than the level of income at any particular time.

From Dimensions to Indicators

The various dimensions are then broken down into variables, some of which will be retained as indicators, either because they seem to be particularly pertinent or because they are easier to measure. While the selection of indicators is often based on an assessment of observation and measurement constraints, it does nevertheless always include theoretical elements. For example, again on poverty, there is a theoretical question which conditions the nature of the income indicator, i.e. is poverty an absolute or relative reality? In other words, should people be considered poor if they do not have the minimum income to cover needs considered to be essential, or if they have considerably less income than other people? In the first case, the poverty threshold will be arrived at by calculating the amounts necessary to cover the needs considered to be essential, which will have to be previously defined. In the second case, measuring the phenomenon will require to set a reference level (distribution mean or median), a spread compared to it (40%, 50%, 60%) and the appropriate scale (household or individual).

From Indicators to Measurements

Once indicators are defined, they must be measured. Then must be decided the level of precision, accuracy, spatial and temporal scale as well as which units are to be used. More often than not, indicators do not have the same degree of precision and are not measured with similar units, which of course complicate the process of aggregation of measurements into a synthetic indicator. For example, the concept of social status, operated by indicators such as length of schooling, level of education, income and type of job, is a mix of purely quantitative (income), semi-quantitative (level of education) and purely qualitative data (job). As a result, it is often necessary to bring down units and measurement scales to the most elementary and least demanding levels, with all that this implies in terms of loss of information.

From Measurements to Index

The last operation—an essential one in the context of putting a scientific concept to the empirical test—is to aggregate the various indicators into a synthetic indicator. When testing a scientific hypothesis (the situation being different in the case of social indicators) only the synthetic indicator is considered significant; basic indicators being meaningless individually; they are just pieces of a puzzle of which only the whole is significant. But to become aggregated, indicators must be capable of expression in a common unit. This is obviously the case for monetary indicators such as GDP, the price index, etc. But if there is no natural common unit such as currency, the different indicators have to be standardised.

Standardisation

There are several possibilities for standardising, none of them entirely satisfactory.

Statistical Standardisation

Statistical standardisation consists in expressing all the values as standard deviations, after having transformed the variables so that their mean is equal to zero. This type of standardisation is done before a great many statistical modelling exercises but is unfortunately inapplicable in the context of social indicators because each new observation involves a new calculation of the mean followed by a new standardisation.

Empirical Standardisation

To be more precise, we should put empirical standardisation in the plural since various techniques can be used. One of the more common ones consists in using as a base for calculation a base-year (for example the year when the statistical survey began) and expressing all the subsequent values as a percentage of variation from the initial value. This approach is useful for an analysis in terms of progress or regression from an initial situation. Another method consists in attributing a 0 value (minimum) to the observation considered as the worst case and 1 (or 10 or 100) to the one corresponding to the best score (maximum). All the intermediate values are then calculated according to the following formula:

$$Y = X - Min/(Max - Min)$$

It is so as to remain within the limits of a scale ranging from 0 to 1 (or 10, 100). The main problem with this type of standardisation is the variability of the minimum and maximum boundaries. If a new observation spills over, either at the top or the bottom of the scale of observations up to that time, all the variables need to be re-standardised, failing which any new observation will be outside the range.

Axiological Standardisation

The process is identical to empirical standardisation with the min and max boundaries, except that the boundaries are not dictated by the data base (observed values) but are chosen with reference to the context of action or evaluation. The situation from which there needs to be differentiation is given the value 0, and the situation which is viewed as ideal (which may or may not correspond to a strategic objective) is given the value 1.

Mathematical Standardisation

Mathematical standardisation consists in applying a mathematical transform (function) to data so that they remain between a lower and a higher boundary (e.g. -1 and +1 or 0 and 1). The logistical and hyperbolic tangent functions are those most frequently used. However, such manipulations are not recommended for social indicators, firstly because they distort to a certain extent the original distribution, but mainly because they lack transparency for a non-professional user. Clearly, the choice of a method and the maximum and minimum boundaries used for standardisation are not without consequence as regards the interpretation and the use of indicators. Bouyssou et al. give several examples of distortion as a result of minute differences in the choice of one or the other baseline values. Take for example the Human Development Index: one of the three components is life expectancy at birth, the observed values of which are standardised with a lower boundary set at

25 years and an upper limit at 85. What would be the result if instead of using 85 years as the upper limit we were to choose 80? The interval between the maximum and the minimum value would change from 60 to 55, i.e. a 9% reduction. A 55-year life expectancy, instead of being worth 0.50, would be worth 0.545, i.e. 9% more. If the other components of the index did not change, the result would be an increase of 9% in the weight of life expectancy in the calculation of the total. As a consequence, the more or less arbitrary nature of the choice of min and max values, even in the case of empirical standardisation, pleads in favour of the adoption of a normative approach and therefore for maximum values to be chosen so that they effectively correspond to the goals to be arrived at.

Aggregation

Aggregation is the operation consisting in condensing the information contained in each criterion into one single item of information. This supposes that the following questions receive an answer. Should the same weight be given to all the criteria constituting the index? Or should they be given different weights? And if so, how? What is the relationship between the index and the indicators? Is it a sum, a product, or something more complicated?

In practice, both questions usually come down to a dilemma between a simple and a weighted average. The question of weighting is a crucial and distinctly difficult one. It consists in attributing a weight, and therefore a specific value to the various dimensions of the concept. For instance, in the case of a poverty index, it could consist in giving more weight to the material dimension than to the social (isolation, exclusion) or cultural dimensions.

Dimensions and indicators making up an index can be represented in the form of a tree diagram, the concept being the trunk of the tree and each branch representing one of the dimensions, with each branch breaking down into sub-branches ending up with the leaves representing the actual indicators. At each branching out, a weighting can be attributed to the branches arising there, with at the end the leaves to which is attached a weight equal to the product of the coefficients of the sub-branches and the branches from which they arise.

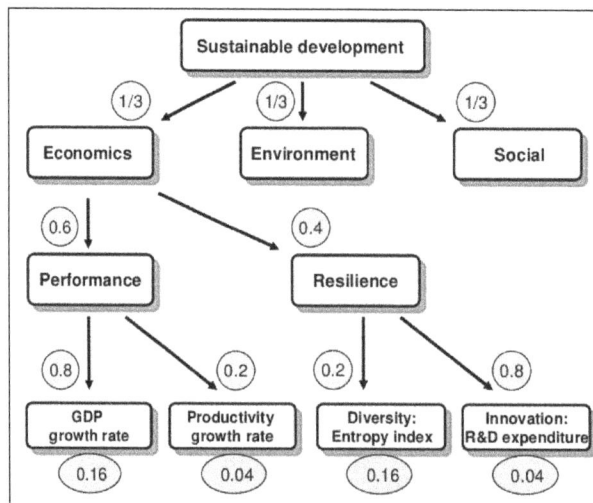

Tree diagram of dimensions and indicators.

Figure above is an example of a tree diagram of this kind where the concept of sustainable development is broken down into three dimensions corresponding to the famous: Economic, Social and Environmental pillars. Only the Economic branch is further developed, with two constituting dimensions, Performance and Resilience. Performance is evaluated with the help of two indicators: two growth rates (GDP and Productivity). The Resilience sub-branch also gives rise to two dimensions: Diversity and Innovation. The cascading weighting process is illustrated by the final weight of each indicator, which is the product of all the previous weights and its own. Thus the GDP growth rate is given a 0.16 weighting, i.e. the product of its own specific weight 0.8, of the 0.6 weight of the "Performance" branch, and the 0.33 weight of the "Economics" branch.

Construction of Indicators and Multi-criteria Decision-making

The hierarchical tree analysis described above is reminiscent of certain methods of multi attribute decision making which use the same kind of decision-tree. As Bouyssou et al rightly remarked, the construction process of indicators is, in fact, a multi-criteria or multi-attribute decision problem. In essence, it is composed of:

- $C = C_1 ... C_n$, a set of objectives to arrive at or of criteria to be taken into consideration (for example, for purchasing a car: price, safety, fuel consumption, etc.);

- $A = A_1 ... A_m$, a finite set of alternative means to arrive at these objectives or meet these criteria (the different car models);

- $W = W_1 ... W_n$, a set (which may be empty) of weightings of criteria C.

such as,

$$\sum_{i=1}^{n} wi = 0$$

The decision consists in ordering the m alternatives on the basis, either of a single criterion made up of the aggregation of the n objectives (or criteria), or the different criteria plurally acquired (the multi-criteria approach), all of which serves to evidence the alternative which is the closest to the desired goal.

The approach consists in filling in an alternatives/criteria matrix made up of the values given by the decision-maker to each alternative as it relates to each criterion. The matrix is then interpreted so as to obtain a classification of the various alternatives and identifying the one which is the closest to satisfying the requirements. In the case of a monocriterion (or aggregative) approach, the entire matrix will be synthesised into a vector comprising only one value per alternative. In a multicriterion approach, although the entire matrix may not be considered, there will at least be consideration of a number of criteria greater than 1.

Let us now take the case of an NGO wishing to set up its international headquarters in the best-performing country as regards sustainable development. It will start by selecting a series of economic, social and environmental indicators, collect the relevant data over a certain number of years and examine the performances of the various countries in terms of sustainable development. Depending on such performances, it will be able to determine the ideal location for its headquarters. This is in fact a decision-making problem where the criteria to consider are indicators which may be weighted and aggregated or, at the very least, synthesised so as to be able to classify the alternatives (the countries).

Two consequences arise out of the similarity of situations: on the one hand, the methods and tools developed as part of the aid to decision-making can equally apply to both the weighting and the aggregation of criteria for sustainable development and therefore to the indicators which account for it; on the other hand, were no aggregated indicator to be produced, this would be comparable to deciding not to classify the various alternatives. Clearly, in the case of sustainable development indicators, this is a matter for collective decision, therefore of social choice, and it is in these terms that it must be considered.

Weighting

While standardisation and aggregation methods raise serious theoretical and practical difficulties, it is mostly as regards weighting that the main scientific challenges and democratic issues arise. As B. Perret rightly remarked, "The intrinsic theoretical weakness of synthetic indicators is obvious (a rational justification of the weightings used is difficult)". On what basis and using what procedure should the decision be made, for example, to give the economic pillar a 45% weighting, 35% to the social pillar and 20% to the environmental one? Does this not suppose that the crucial question of possible substitutions between various kinds of assets has been solved? The temptation is strong to take such weightings for substitution rates (a loss of one point in the environmental pillar can be offset by a gain of 20/45 (0.44) point in the economic pillar, for example). The certain aggregation conventions (called "non compensatory") can limit the risk of erroneous interpretation, but nevertheless current scientific knowledge cannot in itself justify any weighting structure applied to such different sectors.

Is such an exercise actually meaningful? Are we not confronted with an insurmountable obstacle because of the intrinsic incommensurability of the sectors we are trying to compare? On this subject, Martinez-Alier et al., in the context of multicriteria and multi-actor decision-making methods, speak of weak comparability when there is no common basis for comparison with which to rank the various alternatives without leading to a conflict in values. The criteria considered would therefore be incommensurable, for technical reasons, because the real systems are too complex, and/or social reasons, because of the multiplicity of legitimate value systems within society. Why not then abandon the idea of weighting altogether? This is exactly what certain multicriteria and multi-decider analysis techniques do, e.g. the Electre IV method. And yet, every decision, be it individual or collective, contains some arbitrary options, more often than not subconscious and implicit, such as choosing between today or tomorrow, us or them, economic growth or protecting the environment, employment or quality of life, etc. In the realm of public policy, weighting is therefore in the last analysis, the reflection or the echo of the relative power of the various social groups. But the requirements of sustainable development in fact imply an evaluation of these arbitrary choices, in the context of democratic debate and in the light of ethical and scientific criteria. And it is precisely because it forces us to put on the political agenda an evaluation of these choices and weights, which are the components of life in society, that constructing synthetic indices for sustainable development is necessary. It is only through democratic debate between randomly selected citizens independent of any pressure group, that abides by proven procedures in mechanisms such as citizen juries, planning units and hybrid forums, that real collective intent can be expressed. Existing consultative bodies are, from this point of view, the worst of all solutions, as J.-J. Rousseau had long ago stated:

> "If, when the people, being furnished with adequate information, held its deliberations, the citizens had no communication one with another, the grand total of the small differences would always give the general will, and the decision would always be good. But when factions arise, and partial associations are formed at the expense of the great association, the will of each of these associations becomes general in relation to its members, while it remains particular in relation to the State: it may then be said that there are no longer as many votes as there are men, but only as many as there are associations."

Indicators for Whom?

The reasons which disqualify the synthetic index option and argue in favour of the scoreboard are impossible to understand if the user for which the information is provided is not specified. For example, the argument given by Baneth, in opposition to synthetic indices, which reads: "A pilot flies an aircraft using data supplied by a large number of instruments and that data cannot be summed up in a single indicator", is only acceptable if you consider that only pilots, not passengers, need indicators. The aircraft metaphor is irrelevant because the difference between it and a human group or society, is that the passengers of an aircraft are all going to the same destination and all want to get there as safely and comfortably as possible. As a result, once aboard, their only concern is how far they are from their point of arrival and how much time will be needed to get there. This information is in fact displayed on video screens where flight is symbolised by the picture of an airplane moving across a map. In a human society, things are very different. All its citizens do not have, a priori, the same destination and perhaps most of them do not even know where they are going. Before even thinking about steering the social aircraft, its pilots must try to get everyone to agree on where they are headed. This is exactly where indicators for sustainable development come into play.

On closer inspection, indicators can be used for as many social appropriations and purposes as there are policy concepts and, in a democratic society, as there are concepts of democracy. The "aggregative" model in liberal democracies sees the political process as a simple choice, by voting, between a priori preferences which were generated before the electoral process. The model is the market, not the forum. Following this view, there is no common good except if it relates to the least conflictual of the possible specific concepts of good or of the good life. In such a context, social indicators would have but a small role to play in a situation where the members of a political system do not need them to verify that decisions taken by the people in charge are in their best interests. They have personal indicators they can use for that purpose: their income, their employment, their pension schemes, their environment, etc.

But there is another model for democracies, the "deliberative" model, in which the political process exists precisely for the purpose of creating a common vision of what is good or just. The vote itself is less important than the deliberative process which is the source of decisional legitimacy, more so than voting or negotiation between parties each seeking to defend their private interests. It is deliberation which makes it possible to transform "pre-reflective" preferences, established ex ante, into ex post reflective preferences, capable of transcending personal opinions and taking the common good into consideration. While in aggregative democracies (the market), preferences are a given and intangible, in deliberative democracies (the forum), they are designed and constructed through rational argumentation during the process of developing a general will. Social indicators then have a much more important role to play, in so far as they can contribute to the construction of a common definition of the situation and to prior agreement on the facts.

The type of addressee for whom the information is mainly intended is what differentiates the two historical traditions from which current social indicators stem. This is the essential difference between administrative statistics and social indicators. The former are a governmental discipline, implemented by the administration in the service and at the behest of central government. Their primary objective is to inform the authorities (and only them) of the state of society. It is not, for that matter, by pure chance that the emergence of statistics came to be associated with the name of Machiavelli.

Social indicators, however, developed along very different lines. Their purpose is not so much to inform government—even though officially reports are addressed to the government—as to allow civil society to evaluate public policies (and in the last resort, government action) and beyond that, evaluate society's entire development. Unlike official statistics, social indicators are meant to be an instrument of democratic evaluation just as much as a management tool in the hands of the authorities alone. The fate of the French Department of Statistics, the Bureau de Statistiques, is an example of the tension which can build up between the two approaches. It was created in 1796, as a division of the Interior Ministry and in 1800-1801 it completed a considerable body of work collecting data involving the use of questionnaires addressed to regional officials (Préfets), on the basis of which it published a large number of monographs on the state of the Nation. Its overriding objective was to inform citizens and reinforce democracy, rather than satisfying administrative requirements. This was so true that Napoleon, whose sole concern was the availability of the information required for levying taxes and organising conscription, put an end to its activities in 1811. The Bureau des Statistiques monographs were therefore an early kind of social reporting insofar as they aimed more at enriching political debate and informing civil society than contributing to the management of public affairs.

Depending on who they are addressed to and for what purpose, when they are part of the democratic process, indicators can serve to discharge one or several of the following functions. They can be an information basis for political decision-making (internal use); in which case we are dealing with traditional statistics: counting, censuses. They can serve to evaluate, internally and/or externally; this is the social indicator approach. They can also be components of the collective definition of a common world, or even of a common good (goals to arrive at, standards to be maintained) and of the means to achieve it (measurement of well-being).

While the first two uses are well known and amply documented, this is far from being the case for the third which has been almost entirely ignored by political philosophy. And yet, we believe it to be essential, particularly as regards sustainable development. There is however a notable exception to this lack of interest in the role of statistical information in the democratic process: the analysis of the role of social enquiry in relation to politics proposed by John Dewey. For Dewey, the public is what is constituted by the awareness of the fact that certain transactions or private activities can generate consequences which affect those who are external to those transactions. Today we would say that the public is born of an awareness of negative externalities. In other words: "The public consists of all those who are affected by the indirect consequences of transactions to such an extent that it is deemed necessary to have those consequences systematically cared for".

Transaction or actions whose consequences affect groups or individuals other than those directly involved thereby belong to the public domain and are the subject of regulation and control. However, as soon as they are no longer considered to be generating indirect consequences, certain activities which were once part of the public domain can return to the private sector. For example, religious rites and beliefs passed from the public to the private domain when the members of a social community ceased to believe that the consequences of individual piety or impiety could have an effect on the community.

The existence of externalities is not sufficient in itself for a public to be constituted; they must also be perceived and understood. According to Dewey, one of the major political problems of the age of technology is that the consequences of certain individual or group behaviours are so diffuse and remote in time that it is no longer possible to perceive them without recourse to what he calls social enquiry, i.e. scientific investigation of a social nature. We are of the opinion that indicators may acquire their full democratic legitimacy in the context of this social enquiry which is essential for the constitution of an appropriate public.

There may, however, be some mismatch between political and public organisation. While a public state always give rise to some kind of political organisation, it may become inadequate because of the emergence of new publics who may then find themselves deprived of any suitable political organisation. In the preface to the second edition of his book, Dewey considered that relations between nations were in the process of acquiring the properties which constitute a public and that, for that very reason, they needed some kind of specific political organisation which they were lacking at the time.

To counteract and control the undesirable consequences of certain activities, the public creates its own political organisation made up of officials and civil servants designated for that purpose. In a democratic organisation based on the right to vote, every person becomes—because he is a member of the electorate—a public official. Therefore, voting is supposed to serve the public interest and not that person's private interests. Of course, remarks Dewey, "He may fail, in effort to represent the interest entrusted to him. But in this respect he does not differ from those explicitly designated public officials who have also been known to betray the interest committed to them instead of faithfully representing it." This language shows clearly that Dewey rejects an aggregative vision of democracy and is so much in favour of the deliberative perspective that he considers that using voting rights to serve personal interests is a perversion of democracy.

Publics are born, assert themselves and disappear as a result of external conditions such that activities which were once charged with consequence lose that quality while other activities emerge, the effects of which turn out to be "stable, uniform, recurrent and irreparable". Alterations in material conditions (technologies in the main) play a major role in such changes. In Dewey's view, the technological changes he was witness to were radically disrupting the situation: "The machine age has so enormously expanded, multiplied, intensified and complicated the scope of the indirect consequences, has formed such immense and consolidated unions in action, on an impersonal rather than a community basis, that the resultant public cannot identify and distinguish itself."

The changes that have occurred since Dewey wrote these lines have only confirmed his intuition. The quest for sustainable development itself was born of growing discomfort in the face of the hitherto unsuspected magnitude of the long term effects of transactions and economic behaviours? And is it not scientific developments (the social enquiry) which have made us aware that some of our behaviours may affect durably and irreversibly human beings very far away from us in space and in time (future generations)? This explains why certain behaviours which were strictly confined to the private sphere are beginning to enter the public sphere. One example is the management of household waste in which Governments are taking an ever increasing interest by way of regulation, tax incentives, etc.

Very obviously, we are far from being able to appreciate fully the indirect environmental and socio-political consequences of our production and consumption patterns. The public which is building up in relation to these issues still needs structuring; it must find a suitable political organisation for itself and seek out, with the help of this social enquiry process in which indicators of sustainable development are an essential cog, the information needed for action.

References

- "Transforming our world: the 2030 Agenda for Sustainable Development". United Nations – Sustainable Development knowledge platform. Retrieved 23 August 2015

- The-concept-of-sustainable-development: e-ir.info, Retrieved 23 August, 2019

- Thewissen, Stefan; Ncube, Mthuli; Roser, Max; Sterck, Olivier (1 February 2018). "Allocation of development assistance for health: is the predominance of national income justified?". Health Policy and Planning. 33 (suppl_1): i14–i23. Doi:10.1093/heapol/czw173. ISSN 0268-1080. PMC 5886300. PMID 2941

- Zireb-2018-0005.xml, zireb-2018-0005, zireb, view: degruyter.com, Retrieved 24 January, 2019

- Kellogg, Diane M. (2017). "The Global Sanitation Crisis: A Role for Business". Beyond the bottom line: integrating sustainability into business and management practice. Gudić, Milenko, Tan, Tay Keong, Flynn, Patricia M. Saltaire, UK: Greenleaf Publishing. ISBN 9781783533275. OCLC 982187046

Environment and Sustainable Development

The responsible interaction with the environment for the purpose of avoiding the depletion or degradation of natural resources and allowing for the long-term environmental quality is known as environmental sustainability. This chapter has been carefully written to provide an easy understanding of the different environmental issues as well as climate change mitigation.

The goal of environmental sustainability is to conserve natural resources and to develop alternate sources of power while reducing pollution and harm to the environment. For environmental sustainability, the state of the future – as measured in 50, 100 and 1,000 years is the guiding principle. Many of the projects that are rooted in environmental sustainability will involve replanting forests, preserving wetlands and protecting natural areas from resource harvesting. The biggest criticism of environmental sustainability initiatives is that their priorities can be at odds with the needs of a growing industrialized society.

Development is not simply about the interactions between human groups; it also involves the natural environment. So, from another point of view, development is about the conversion of natural resources into cultural resources. This conversion has taken place throughout the history of human societies, although the process has generally increased in pace and complexity with time. If we use a system diagram to illustrate – in very general terms – what an economy does, we see that the basic function of an economy is to convert natural resources (in the forms of raw materials and energy) into products and services that are useful to humans. Inevitably, because conversion processes are never totally efficient, some waste is produced which is usually discarded into the environment as various forms of pollution. Therefore, the environment is both a source and a sink in relation to economic processes: it is a source of raw materials and energy and a sink for pollution.

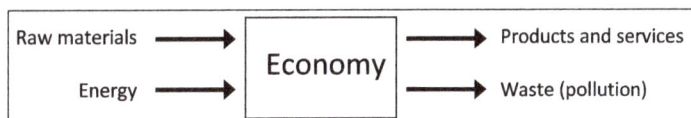

A representation of a generalised economy.

Resources, Energy and Waste

An example of this type of conversion would be the extraction of crude oil from the North Sea, its fractionation and distillation in oil refineries, and its conversion to petroleum or diesel. In turn, those products (petrol and diesel) are converted – through combustion processes – into useful work (such as transportation) whilst the waste products are released into the atmosphere as greenhouse gases (such as carbon dioxide). If we add together all of the conversion processes that occur, for instance, in a given country, we would have a sense of the total input and output of that national economy. This could be expressed in terms of the total natural resources and energy consumed, the total products and services created and the total pollution generated. (In fact, the total value of the finished products and services created in a given country is expressed using a widely-used measure, the Gross Domestic Product, or GDP). If we wanted to increase the creation of products and services, in a given economy, we would require more natural resources and energy, and we would also generate more pollution as a by-product.

Economic Growth

From this point of view, development means an increase in the size or pace of the economy such that more products and services are produced. Conventionally, a common assumption has been that, if an economy generates more products and services, then humans will enjoy a higher standard of living. The aim of many conventional approaches to development has been to increase the size of the economy (economic growth) in order to increase the output of products and services. Of course, without any change in the fundamental economic processes involved, the production of more products and services will inevitably require more raw materials and energy, and will generate more waste. In a system diagram, this would simply be represented by greater flows of materials and energy through the central box, the economy.

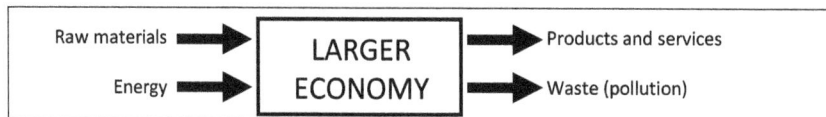

A representation of economic growth.

Unsustainable Development

Effects of Greater Throughput

The fact that economic growth means an increase in the throughput of an economy raises several issues:

- Whilst some raw materials (such as air) are ubiquitous and others are readily available, many raw materials are scarce and their availability cannot be guaranteed indefinitely.

- Similarly, some sources of energy (such as the wind) are renewable and freely available, whilst others (such as fossil fuels) are non-renewable and finite.

- Most pollution sinks have a limited capacity to absorb the waste by-products of economic processes.

- In affluent societies, problems of overconsumption have emerged and questions are now being raised about the extent to which the acquisition of additional products and services actually improves well-being in those societies.

These considerations hint at an important idea: that development can be "unsustainable" insofar as it cannot continue indefinitely if economic growth exhausts the available supplies of raw materials, the sources of energy or the pollution sinks. But suppose economic growth does reach the limits of raw material supplies, energy supplies or waste assimilation capacity: what happens then? The following example illustrates what can happen in such a situation.

Tragedy of Easter Island

Easter Island Mystery

Easter Island is one of the remotest inhabited places on Earth. It is a small island (around 400 square kilometres) in the Pacific Ocean, approximately 2000 kilometres from the nearest habitable land (Pitcairn Island). Despite the small size and remote location of the island, at the peak of its society, it had a human population of 7000 people. Yet even that small population was to place demands upon the natural environment that could not be sustained. At the time of first contact with Europeans in 1722, around 3000 people were found on the island in desperate conditions. Subsequently, the population continued to decline and their living conditions worsened. In 1877, Peruvians removed and enslaved the remaining population, with the exception of 110 elderly people and children. Eventually, the island was annexed by Chile and was leased to a British company for sheep grazing, with the few remaining inhabitants being confined to a single small village.

The mystery that faced the first European visitors was that, despite the appalling conditions they found on the island, there was also evidence of a once-flourishing and advanced society. Over 600 huge stone statues (averaging over 6 metres in height) were found on the island. The task of carving, transporting and erecting the statues was a complex one and was undeniably beyond the capacity of the poverty-stricken inhabitants of Easter Island in 1722. Indeed, given the limited resources of Easter Island, the society that constructed the statues must have been one of the most advanced in the world for the technology they had available. So what had happened to the statue-makers? Modern archaeological techniques revealed that the advanced society that constructed the statues collapsed because the development that occurred on the island placed immense demands – that could not be sustained — on the natural environment of the island. This makes the history of Easter Island a powerful example of the dependence of human societies on their natural environment — and of the consequences of irreversibly damaging that environment.

Collapse of a Society

The colonisation of Easter Island began in the fifth century by Polynesians during a major phase of exploration and settlement across the Pacific Ocean. The first settlers would have found a volcanic landscape with adequate soils but poor drainage and few fresh water supplies. Due to its remote location, the island had few plant and animal species and the surrounding waters contained few fish. Hence, the settlers relied on a very limited range of plants and animals for their subsistence: their diet would have consisted mainly of chicken and sweet potatoes. As the population of Easter Island grew, familiar forms of Polynesian social organisation were introduced. Ceremonial activities — including elaborate rituals and monument construction — became a major part of the social life of the islanders. The growth of the population was accompanied by increased competition between clan groups. In turn, the outcome of that competition was the creation of one of the most complex Polynesian societies, the construction of hundreds of ceremonial centres with large stone platforms, and the carving of the stone statues. It was at this point that the society suddenly collapsed, leaving more than half of the statues partially completed. Why did this collapse occur?

The cause of the collapse of the society was the extensive environmental degradation resulting from the deforestation of the entire island. The Europeans who visited Easter Island in 1722 found the island completely denuded of trees (with the exception of some isolated trees at the bottom of a deep volcanic crater). Yet scientific analysis indicates that, at the time of initial settlement, the island would have been densely vegetated with large areas of woodland. Those trees were cleared by the growing islander population to provide clearings for agriculture, fuel for cooking and warmth, and a source of material for the construction of housing and canoes. Above all, vast quantities of wood were needed to transport the enormous statues to the ceremonial sites around the island; that task was accomplished using an elaborate system of wooden tracks. As a result of the excessive demand for wood, the island was almost entirely deforested by 1600 and the construction of statues halted. By that time, other effects of deforestation would have been apparent. House-building became impossible and people resorted to living in caves, stone shelters or reed huts. Fishing – that previously used nets made from bark — became more difficult. The construction of canoes became impossible and the population was then unable to escape the island. The removal of trees caused soil erosion, the leaching of nutrients and the decline of crop yields. The combined effect of those changes meant that the population could no longer be supported on a shrinking resource base, and a rapid decline in numbers occurred. After 1600, the remnant society regressed to very primitive living conditions and continued to decline until its eventual disappearance.

An Avoidable Tragedy

The tragedy of Easter Island is that — in a remote and unlikely location — the original inhabitants had flourished and created one of the most advanced societies in the world for the technology available. Their achievements could have been celebrated as a triumph of human ingenuity and capacity to adapt and survive. Ultimately, however, competition between clan groups and the intensity of human demands on the natural resource base exceeded the capacity of the environment. Following abrupt environmental decline, the society collapsed and a substantially reduced population reverted to very poor living conditions. The Easter Islanders must have been aware that they were entirely dependent upon their extremely limited resource base. Environmental changes on the island — especially deforestation — must also have been readily apparent. Nevertheless, the islanders were unable to prevent the destruction of their resource base; instead, key resources were depleted until they were totally exhausted. Archaeological evidence suggests that, instead of prompting careful management of the remaining resources, competition between clans (and the use of timber) intensified as the environmental crisis became more acute. The history of Easter Island suggests that the response of the islanders to their deepening environmental crisis was not one of re-evaluation and restraint, but was desperate and chaotic, and it resulted in the fatal destruction of their life support system.

The Lessons of Easter Island

Development Depends on the Environment

The example of Easter Island, whilst tragic, is useful for illustrating several key points about the relationship between environment and development. There is an intimate relationship between environment and development. Development – understood, in this example, as the increasing use of natural resources by humans for their economic, social, and cultural activities – cannot occur independently of the environment that provides resources and assimilates pollution. (It is worth pointing out that this relationship is certainly not reciprocal; whilst development depends heavily on the environment, the environment does not require development – or human existence – at all). In this example, as in many other cases, the nature of the relationship between environment and development is central to interpretations of whether or not development is sustainable.

In the example of Easter Island, we can conclude that the expansion of human activities on the island was unsustainable because the relationship between environment and development was characterised by the over-exploitation of natural resources, even in a situation where subsistence was already marginal, together with a complete disregard for the warning signs and consequences of environmental degradation.

A Critical Trade-off

On Easter Island, the relationship between humans and their environment was such that a trade-off between environment and development occurred. In other words, natural resources (trees and soil) were progressively exchanged, by the islanders, for a range of economic, social, and cultural benefits (ceremonial activities, stone platforms, statues, dominance, power, and wealth). Such a trade-off becomes inevitable if development conflicts with the need for environmental protection. As a result of the trade-off between environment and development, the Easter Islanders did not leave a sufficient resource base for future generations. In the language of sustainable development, there was no intergenerational equity. Each current generation failed to protect the resources that would be needed by its descendents. Once the deforestation of the island had reached a critical point, future generations were left without the resources they needed to maintain an equivalent way of life. That failure to maintain the resource base for future generations inevitably set up conflict between islanders and made the challenge faced by each successive generation more difficult to surmount.

Above all, the example of Easter Island illustrates the imperative for human societies to live within the capacity of their natural resource base. If excessive demands are placed upon the natural resource base (through deliberate or inadvertent exploitation or mismanagement of the environment), then both natural processes and human activities are bound to decline – perhaps with catastrophic consequences.

A Metaphor for Global Development

The Easter Island example can be regarded as a metaphor for global development. Like Easter Island at the time of its first inhabitants, the Earth has limited resources to support human societies and their myriad demands. Like the stranded islanders, the inhabitants of Earth have no realistic means of escape. Human existence depends, ultimately, upon the continued availability of the Earth's natural resources that support life. In general, over the period of human existence (around 2 million years), human societies have been successful in obtaining food and in extracting natural resources, with the result that growing populations – and increasingly complex, advanced societies – have been sustained. But what about the critical relationship that indicates whether or not development is sustainable: the relationship between environment and development? Have modern societies been more successful than the Easter Islanders in living in a way that does not exhaust the available natural resources? Have we fallen – or are we falling – into the same trap as the Easter Islanders: that of fatally damaging our life support system?

ENVIRONMENTAL ISSUE

Environmental issues are harmful effects of human activity on the biophysical environment. Environmental protection is a practice of protecting the natural environment on individual, organizational or governmental levels, for the benefit of both the environment and humans. Environmentalism, a social and environmental movement, addresses environmental issues through advocacy, education and activism.

The carbon dioxide equivalent of greenhouse gases (GHG) in the atmosphere has already exceeded over 9000 parts per million (NOAA) (with total "long-term" GHG exceeding 455 parts per million) (Intergovernmental Panel on Climate Report). The amount of greenhouse gas in the atmosphere is possibly above the threshold that can potentially cause climate change. The UN Office for the Coordination of Humanitarian Affairs (OCHA) has stated "Climate change is not just a distant future threat. It is the main driver behind rising humanitarian needs and we are seeing its impact. The number of people affected and the damages inflicted by extreme weather has been unprecedented." Further, OCHA has stated:

> Climate disasters are on the rise. Around 70 percent of disasters are now climate related – up from around 50 percent from two decades ago.

> These disasters take a heavier human toll and come with a higher price tag. In the last decade, 2.4 billion people were affected by climate related disasters, compared to 1.7 billion in the previous decade. The cost of responding to disasters has risen tenfold between 1992 and 2008.

> Destructive sudden heavy rains, intense tropical storms, repeated flooding and droughts are likely to increase, as will the vulnerability of local communities in the absence of strong concerted action.

Environment destruction caused by humans is a global problem, and this is a problem that is on going every day. By year 2050, the global human population is expected to grow by 2 billion people, thereby reaching a level of 9.6 billion people (Living Blue Planet 24). The human effects on Earth can be seen in many different ways. A main one is the temperature rise, and according to the report "Our Changing Climate", the global warming that has been going on for the past 50 years is primarily due to human activities Walsh, et al. 20. Since 1895, the U.S. average temperature has increased from 1.3 °F to 1.9 °F, with most of the increase taken place since around year 1970 Walsh, et al. 20.

Types

Major current environmental issues may include climate change, pollution, environmental degradation, and resource depletion etc. The conservation movement lobbies for protection of endangered species and protection of any ecologically valuable natural areas, genetically modified foods and global warming.

Scientific Grounding

The level of understanding of Earth has increased markedly in recent times through science especially with the application of the scientific method. Environmental science is now a multi-disciplinary academic study taught and researched at many universities. This is used as a basis for addressing environmental issues.

Large amounts of data have been gathered and these are collated into reports, of which a common type is the State of the Environment publications. A recent major report was the Millennium Ecosystem Assessment, with input from 1200 scientists and released in 2005, which showed the high level of impact that humans are having on ecosystem services.

Organizations

Environmental issues are addressed at a regional, national or international level by government organizations.

The largest international agency, set up in 1972, is the United Nations Environment Programme. The International Union for Conservation of Nature brings together 83 states, 108 government agencies, 766 Non-governmental organizations and 81 international organizations and about 10,000 experts and scientists from countries around the world. International non-governmental organizations include Greenpeace, Friends of the Earth and World Wide Fund for Nature. Governments enact environmental policy and enforce environmental law and this is done to differing degrees around the world.

Solutions

Sustainability is the key to prevent or reduce the effect of environmental issues. There is now clear scientific evidence that humanity is living unsustainably, and that an unprecedented collective effort is needed to return human use of natural resources to within sustainable limits. For humans to live sustainably, the Earth's natural resources must be used at a rate at which they can be replenished (and by limiting global warming).

Concerns for the environment have prompted the formation of green parties, political parties that seek to address environmental issues. Initially, these were formed in Australia, New Zealand and Germany but are now present in many other countries.

CONSERVATION

Conservation's goals include protecting species from extinction, maintaining and restoring habitats, enhancing ecosystem services and protecting biological diversity. A range of values underlie conservation, which can be guided by biocentrism, anthropocentrism, ecocentrism and sentientism. There has recently been a movement towards evidence-based conservation which calls for greater use of scientific evidence to improve the effectiveness of consecration efforts.

Conservation goals include conserve habitat, preventing deforestation, halting species extinction, reducing overfishing and mitigating climate change. Different philosophical outlooks guide conservationists towards different goals.

The principal value underlying many expressions of the conservation ethic is that the natural world has intrinsic and intangible worth along with utilitarian value – a view carried forward by parts of the scientific conservation movement and some of the older Romantic schools of ecology movement. Philosophers have attached intrinsic value to different aspects of nature, whether this is individual organisms (biocentrism) or ecological wholes such as species or ecosystems (ecoholism).

More utilitarian schools of conservation have an anthropocentric outlook and seek a proper valuation of local and global impacts of human activity upon nature in their effect upon human wellbeing, now and to posterity. How such values are assessed and exchanged among people determines the social, political, and personal restraints and imperatives by which conservation is practiced. This is a view common in the modern environmental movement. There is increasing interest in extending the responsibility for human wellbeing to include the welfare of sentient animals. Branches of conservation ethics focusing on sentient creatures include ecofeminism and compassionate conservation.

In the United States of America, the year 1864 saw the publication of two books which laid the foundation for Romantic and Utilitarian conservation traditions in America. The posthumous publication of Henry David Thoreau's *Walden* established the grandeur of unspoiled nature as a citadel to nourish the spirit of man. From George Perkins Marsh a very different book, *Man and Nature*, later subtitled "The Earth as Modified by Human Action", catalogued his observations of man exhausting and altering the land from which his sustenance derives.

The consumer conservation ethic is sometimes expressed by the *four R's*: " Rethink, Reduce, Recycle, Repair" This social ethic primarily relates to local purchasing, moral purchasing, the sustained, and efficient use of renewable resources, the moderation of destructive use of finite resources, and the prevention of harm to common resources such as air and water quality, the natural functions of a living earth, and cultural values in a built environment.

Practice

The Daintree Rainforest in Queensland, Australia.

Distinct trends exist regarding conservation development. While many countries' efforts to preserve species and their habitats have been government-led, those in the North Western Europe tended to arise out of the middle-class and aristocratic interest in natural history, expressed at the level of the individual and the national, regional or local learned society. Thus countries like

Britain, the Netherlands, Germany, etc. had what we would today term NGOs – in the shape of the RSPB, National Trust and County Naturalists' Trusts (dating back to 1889, 1895 and 1912 respectively) Natuurmonumenten, Provincial Conservation Trusts for each Dutch province, Vogelbescherming, etc. – a long time before there were national parks and national nature reserves. This in part reflects the absence of wilderness areas in heavily cultivated Europe, as well as a longstanding interest in laissez-faire government in some countries, like the UK, leaving it as no coincidence that John Muir, the Scottish-born founder of the National Park movement (and hence of government-sponsored conservation) did his sterling work in the USA, where he was the motor force behind the establishment of such NPs as Yosemite and Yellowstone. Nowadays, officially more than 10 percent of the world is legally protected in some way or the other, and in practice, private fundraising is insufficient to pay for the effective management of so much land with protective status.

Protected areas in developing countries, where probably as many as 70–80 percent of the species of the world live, still enjoy very little effective management and protection. Some countries, such as Mexico, have non-profit civil organizations and landowners dedicated to protecting vast private property, such is the case of Hacienda Chichen's Maya Jungle Reserve and Bird Refuge in Chichen Itza, Yucatán. The Adopt A Ranger Foundation has calculated that worldwide about 140,000 rangers are needed for the protected areas in developing and transition countries. There are no data on how many rangers are employed at the moment, but probably less than half the protected areas in developing and transition countries have any rangers at all and those that have them are at least 50% short. This means that there would be a worldwide ranger deficit of 105,000 rangers in the developing and transition countries.

One of the world's foremost conservationists, Dr. Kenton Miller, stated about the importance of Rangers: "The future of our ecosystem services and our heritage depends upon park rangers. With the rapidity at which the challenges to protected areas are both changing and increasing, there has never been more of a need for well-prepared human capacity to manage. Park rangers are the backbone of park management. They are on the ground. They work on the front line with scientists, visitors, and members of local communities."

Adopt A Ranger, fears that the ranger deficit is the greatest single limiting factor in effectively conserving nature in 75% of the world. Currently, no conservation organization or western country or international organization addresses this problem. Adopt A Ranger has been incorporated to draw worldwide public attention to the most urgent problem that conservation is facing in developing and transition countries: protected areas without field staff. Very specifically, it will contribute to solving the problem by fundraising to finance rangers in the field. It will also help governments in developing and transition countries to assess realistic staffing needs and staffing strategies.

Others, including Survival International, have advocated instead for cooperation with local tribal peoples, who are natural allies of the conservation movement and can provide cost-effective protection.

The terms *conservation* and *preservation* are frequently conflated outside the academic, scientific, and professional kinds of literature. The US National Park Service offers the following explanation of the important ways in which these two terms represent very different conceptions of environmental protection ethics:

"Conservation and preservation are closely linked and may indeed seem to mean the same thing. Both terms involve a degree of protection, but how that protection is carried out is the key difference. Conservation is generally associated with the protection of natural resources, while preservation is associated with the protection of buildings, objects, and landscapes. Put simply, conservation seeks the proper use of nature, while preservation seeks protection of nature from use."

During the environmental movement of the early 20th century, two opposing factions emerged: conservationists and preservationists. Conservationists sought to regulate human use while preservationists sought to eliminate human impact altogether.

Water Conservation

United States 1960 postal stamp advocating water conservation.

Water conservation includes all the policies, strategies and activities to sustainably manage the natural resource of fresh water, to protect the hydrosphere, and to meet the current and future human demand. Population, household size, and growth and affluence all affect how much water is used. Factors such as climate change have increased pressures on natural water resources especially in manufacturing and agricultural irrigation. Many US cities have already implemented policies aimed at water conservation, with much success.

The goals of water conservation efforts include:

- Ensuring availability of water for future generations where the withdrawal of freshwater from an ecosystem does not exceed its natural replacement rate.

- Energy conservation as water pumping, delivery and wastewater treatment facilities consume a significant amount of energy. In some regions of the world over 15% of total electricity consumption is devoted to water management.

- Habitat conservation where minimizing human water use helps to preserve freshwater habitats for local wildlife and migrating waterfowl, but also water quality.

Strategies

The key activities that benefit water conservation (save water) are as follows:

- Any beneficial reduction in water loss, use and waste of resources.

- Avoiding any damage to water quality.

- Improving water management practices that reduce the use or enhance the beneficial use of water.

One strategy in water conservation is rain water harvesting. Digging ponds, lakes, canals, expanding the water reservoir, and installing rain water catching ducts and filtration systems on homes are different methods of harvesting rain water. Many people in many countries keep clean containers so they can boil it and drink it, which is useful to supply water to the needy. Harvested and filtered rain water can be used for toilets, home gardening, lawn irrigation, and small scale agriculture.

Another strategy in water conservation is protecting groundwater resources. When precipitation occurs, some infiltrates the soil and goes underground. Water in this saturation zone is called groundwater. Contamination of groundwater causes the groundwater water supply to not be able to be used as a resource of fresh drinking water and the natural regeneration of contaminated groundwater can take years to replenish. Some examples of potential sources of groundwater contamination include storage tanks, septic systems, uncontrolled hazardous waste, landfills, atmospheric contaminants, chemicals, and road salts. Contamination of groundwater decreases the replenishment of available freshwater so taking preventative measures by protecting groundwater resources from contamination is an important aspect of water conservation.

An additional strategy to water conservation is practicing sustainable methods of utilizing groundwater resources. Groundwater flows due to gravity and eventually discharges into streams. Excess pumping of groundwater leads to a decrease in groundwater levels and if continued it can exhaust the resource. Ground and surface waters are connected and overuse of groundwater can reduce and in extreme examples, diminish the water supply of lakes, rivers, and streams. In coastal regions, over pumping groundwater can increase saltwater intrusion which results in the contamination of groundwater water supply. Sustainable use of groundwater is essential in water conservation.

A fundamental component to water conservation strategy is communication and education outreach of different water programs. Developing communication that educates science to land managers, policy makers, farmers, and the general public is another important strategy utilized in water conservation. Communication of the science of how water systems work is an important aspect when creating a management plan to conserve that system and is often used for ensuring the right management plan to be put into action.

Social Solutions

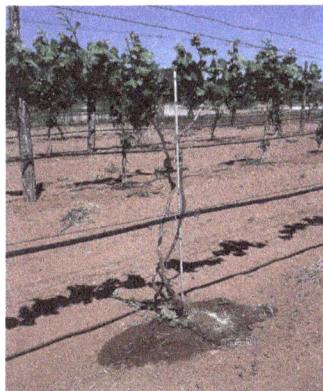

Drip irrigation system in New Mexico.

Water conservation programs involved in social solutions are typically initiated at the local level, by either municipal water utilities or regional governments. Common strategies include public outreach campaigns, tiered water rates (charging progressively higher prices as water use increases), or restrictions on outdoor water use such as lawn watering and car washing. Cities in dry climates often require or encourage the installation of xeriscaping or natural landscaping in new homes to reduce outdoor water usage. Most urban outdoor water use in California is residential, illustrating a reason for outreach to households as well as businesses.

One fundamental conservation goal is universal metering. The prevalence of residential water metering varies significantly worldwide. Recent studies have estimated that water supplies are metered in less than 30% of UK households, and about 61% of urban Canadian homes (as of 2001). Although individual water meters have often been considered impractical in homes with private wells or in multifamily buildings, the U.S. Environmental Protection Agency estimates that metering alone can reduce consumption by 20 to 40 percent. In addition to raising consumer awareness of their water use, metering is also an important way to identify and localize water leakage. Water metering would benefit society, in the long run, it is proven that water metering increases the efficiency of the entire water system, as well as help unnecessary expenses for individuals for years to come. One would be unable to waste water unless they are willing to pay the extra charges, this way the water department would be able to monitor water usage by the public, domestic and manufacturing services.

Some researchers have suggested that water conservation efforts should be primarily directed at farmers, in light of the fact that crop irrigation accounts for 70% of the world's fresh water use. The agricultural sector of most countries is important both economically and politically, and water subsidies are common. Conservation advocates have urged removal of all subsidies to force farmers to grow more water-efficient crops and adopt less wasteful irrigation techniques.

New technology poses a few new options for consumers, features such as full flush and half flush when using a toilet are trying to make a difference in water consumption and waste. Also available are modern shower heads that help reduce wasting water: Old shower heads are said to use 5-10 gallons per minute, while new fixtures available use 2.5 gallons per minute and offer equal water coverage. Another method is to recycle the water of the shower directly, by means a semi-closed system which features a pump and filter. Besides recycling water, it also reuses the heat of the water (which would otherwise be lost).

Household Applications

The Home Water Works website contains useful information on household water conservation. Contrary to the popular view that the most effective way to save water is to curtail water-using behavior (e.g., by taking shorter showers), experts suggest the most efficient way is replacing toilets and retrofitting washers; as demonstrated by two household end use logging studies in the U.S.

Water-saving technology for the home includes:

1. Low-flow shower heads sometimes called energy-efficient shower heads as they also use less energy.

2. Low-flush toilets and composting toilets. These have a dramatic impact in the developed world, as conventional western toilets use large volumes of water.

3. Dual flush toilets created by Caroma includes two buttons or handles to flush different levels of water. Dual flush toilets use up to 67% less water than conventional toilets.

4. Faucet aerators, which break water flow into fine droplets to maintain "wetting effectiveness" while using less water. An additional benefit is that they reduce splashing while washing hands and dishes.

5. Raw water flushing where toilets use sea water or non-purified water.

6. Wastewater reuse or recycling systems, allowing:

   o Reuse of graywater for flushing toilets or watering gardens.

   o Recycling of wastewater through purification at a water treatment plant.

7. Rainwater harvesting.

8. High-efficiency clothes washers.

9. Weather-based irrigation controllers.

10. Garden hose nozzles that shut off the water when it is not being used, instead of letting a hose run.

11. Low flow taps in wash basins.

12. Swimming pool covers that reduce evaporation and can warm pool water to reduce water, energy and chemical costs.

13. Automatic faucet is a water conservation faucet that eliminates water waste at the faucet. It automates the use of faucets without the use of hands.

Commercial Applications

Many water-saving devices (such as low-flush toilets) that are useful in homes can also be useful for business water saving. Other water-saving technology for businesses includes:

1. Waterless urinals.

2. Waterless car washes.

3. Infrared or foot-operated taps, which can save water by using short bursts of water for rinsing in a kitchen or bathroom.

4. Pressurized waterbrooms, which can be used instead of a hose to clean sidewalks.

5. X-ray film processor re-circulation systems.

6. Cooling tower conductivity controllers.

7. Water-saving steam sterilizers, for use in hospitals and health care facilities.

8. Rain water harvesting.

9. Water to Water heat exchangers.

Agricultural Applications

For crop irrigation, optimal water efficiency means minimizing losses due to evaporation, runoff or subsurface drainage while maximizing production. An evaporation pan in combination with specific crop correction factors can be used to determine how much water is needed to satisfy plant requirements. Flood irrigation, the oldest and most common type, is often very uneven in distribution, as parts of a field may receive excess water in order to deliver sufficient quantities to other parts. Overhead irrigation, using center-pivot or lateral-moving sprinklers, has the potential for a much more equal and controlled distribution pattern. Drip irrigation is the most expensive and least-used type, but offers the ability to deliver water to plant roots with minimal losses. However, drip irrigation is increasingly affordable, especially for the home gardener and in light of rising water rates. Using drip irrigation methods can save up to 30,000 gallons of water per year when replacing irrigation systems that spray in all directions. There are also cheap effective methods similar to drip irrigation such as the use of soaking hoses that can even be submerged in the growing medium to eliminate evaporation.

Overhead irrigation, center pivot design.

As changing irrigation systems can be a costly undertaking, conservation efforts often concentrate on maximizing the efficiency of the existing system. This may include chiselling compacted soils, creating furrow dikes to prevent runoff, and using soil moisture and rainfall sensors to optimize irrigation schedules. Usually large gains in efficiency are possible through measurement and more effective management of the existing irrigation system. The 2011 UNEP Green Economy Report notes that "improved soil organic matter from the use of green manures, mulching, and recycling of crop residues and animal manure increases the water holding capacity of soils and their ability to absorb water during torrential rains", which is a way to optimize the use of rainfall and irrigation during dry periods in the season.

Water Reuse

Water shortage has become an increasingly difficult problem to manage. More than 40% of the world's population live in a region where the demand for water exceeds its supply. The imbalance between supply and demand, along with persisting issues such as climate change and population growth, has made water reuse a necessary method for conserving water. There are a variety of methods used in the treatment of waste water to ensure that it is safe to use for irrigation of food crops and drinking water.

Seawater desalination requires more energy than the desalination of fresh water. Despite this, many seawater desalination plants have been built in response to water shortages around the world. This makes it necessary to evaluate the impacts of seawater desalination and to find ways to improve desalination technology. Current research involves the use of experiments to determine the most effective and least energy intensive methods of desalination.

Sand filtration is another method used to treat water. Recent studies show that sand filtration needs further improvements, but it is approaching optimization with its effectiveness at removing pathogens from water. Sand filtration is very effective at removing protozoa and bacteria, but struggles with removing viruses. Large-scale sand filtration facilities also require large surface areas to accommodate them.

The removal of pathogens from recycled water is of high priority because wastewater always contains pathogens capable of infecting humans. The levels of pathogenic viruses have to be reduced to a certain level in order for recycled water to not pose a threat to human populations. Further research is necessary to determine more accurate methods of assessing the level of pathogenic viruses in treated wastewater.

Wasting of Water

Leaking garden hose bib.

Wasting of water (also called "water waste") is the flip side of water conservation and in household applications, it means causing or permitting discharge of water without any practical purpose. Inefficient water use is also considered wasteful. By EPA estimate, household leaks in the U.S. can waste approximately 900 billion gallons (3.4 billion cubic meters) of water annually nationwide. Generally, water management agencies are reluctant or unwilling to give a concrete definition to the somewhat fuzzy concept of water waste. However, definition of water waste is often given in local drought emergency ordinances. One example refers to any acts or omissions, whether willful or negligent, that are "causing or permitting water to leak, discharge, flow or run to waste into any gutter, sanitary sewer, watercourse or public or private storm drain, or to any adjacent property, from any tap, hose, faucet, pipe, sprinkler, pond, pool, waterway, fountain or nozzle." In this example, the city code also clarifies that in the case of "washing", "discharge," "flow" or "run to waste" means that water in excess of that necessary to wash, wet or clean the dirty or dusty object, such as an automobile, sidewalk, or parking area, flows to waste. Water utilities (and other media sources) often provide listings of wasteful water-use practices and prohibitions of wasteful uses. Examples include utilities in San Antonio, Texas. Las Vegas, Nevada, California Water Service company in California, and City of San Diego, California. The City of Palo Alto in California enforces permanent water use restrictions on wasteful practices such as leaks, runoff, irrigating during and immediately after rainfall, and use of potable water when non-potable water is available. Similar restrictions are in effect in the State of Victoria, Australia. Temporary water use bans (also known as "hosepipe bans") are used in England, Scotland, Wales and Northern Ireland.

Strictly speaking, water that is discharged into the sewer, or directly to the environment is not wasted or lost. It remains within the hydrologic cycle and returns to the land surface and surface water bodies as precipitation. However, in many cases, the source of the water is at a significant distance from the return point and may be in a different catchment. The separation between extraction point and return point can represent significant environmental degradation in the watercourse and riparian strip. What is "wasted" is the community's supply of water that was captured, stored, transported and treated to drinking quality standards. Efficient use of water saves the expense of water supply provision and leaves more fresh water in lakes, rivers and aquifers for other users and also for supporting ecosystems. A concept that is closely related to water wasting is "water-use efficiency." Water use is considered inefficient if the same purpose of its use can be accomplished with less water. Technical efficiency derives from engineering practice where it is typically used to describe the ratio of output to input and is useful in comparing various products and processes. For example, one showerhead would be considered more efficient than another if it could accomplish the same purpose (i.e., of showering) by using less water or other inputs (e.g., lower water pressure). However, the technical efficiency concept is not useful in making decisions of investing money (or resources) in water conservation measures unless the inputs and outputs are measured in value terms. This expression of efficiency is referred to as economic efficiency and is incorporated into the concept of water conservation.

CLIMATE CHANGE MITIGATION

Climate change mitigation consists of actions to limit the magnitude or rate of long-term global warming and its related effects. Climate change mitigation generally involves reductions in human (anthropogenic) emissions of greenhouse gases (GHGs). Mitigation may also be achieved by increasing the capacity of carbon sinks, e.g., through reforestation. Mitigation policies can substantially reduce the risks associated with human-induced global warming.

According to the IPCC's 2014 assessment report, "Mitigation is a public good; climate change is a case of the 'tragedy of the commons'. Effective climate change mitigation will not be achieved if each agent (individual, institution or country) acts independently in its own selfish interest, suggesting the need for collective action. Some adaptation actions, on the other hand, have characteristics of a private good as benefits of actions may accrue more directly to the individuals, regions, or countries that undertake them, at least in the short term. Nevertheless, financing such adaptive activities remains an issue, particularly for poor individuals and countries."

Examples of mitigation include reducing energy demand by increasing energy efficiency, phasing out fossil fuels by switching to low-carbon energy sources, and removing carbon dioxide from Earth's atmosphere, for example, through improved building insulation. Another approach to climate change mitigation is climate engineering.

Most countries are parties to the United Nations Framework Convention on Climate Change (UN-FCCC). The ultimate objective of the UNFCCC is to stabilize atmospheric concentrations of GHGs at a level that would prevent dangerous human interference of the climate system. Scientific analysis can provide information on the impacts of climate change, but deciding which impacts are dangerous requires value judgments.

In 2010, Parties to the UNFCCC agreed that future global warming should be limited to below 2.0 °C (3.6 °F) relative to the pre-industrial level. With the Paris Agreement of 2015 this was confirmed, but was revised with a new target laying down "parties will do the best" to achieve warming below 1.5 °C. The current trajectory of global greenhouse gas emissions does not appear to be consistent with limiting global warming to below 1.5 or 2 °C. Other mitigation policies have been proposed, some of which are more stringent or modest than the 2 °C limit. In 2019, after 2 years of research, scientists from Australia, and Germany presented the "One Earth Climate Model" showing how temperature increase can be limited to 1.5 °C for 1.7 trillion dollars a year.

Greenhouse Gas Concentrations and Stabilization

One of the issues often discussed in relation to climate change mitigation is the stabilization of greenhouse gas concentrations in the atmosphere. The United Nations Framework Convention on Climate Change (UNFCCC) has the ultimate objective of preventing "dangerous" anthropogenic (i.e., human) interference of the climate system. As is stated in Article 2 of the Convention, this requires that greenhouse gas (GHG) concentrations are stabilized in the atmosphere at a level where ecosystems can adapt naturally to climate change, food production is not threatened, and economic development can proceed in a sustainable fashion.

There are a number of anthropogenic greenhouse gases. These include carbon dioxide (chemical formula: CO_2), methane (CH_4), nitrous oxide (N_2O), and a group of gases referred to as halocarbons. Another greenhouse gas, water vapor, has also risen as an indirect result of human activities. The emissions reductions necessary to stabilize the atmospheric concentrations of these gases varies. CO_2 is the most important of the anthropogenic greenhouse gases.

There is a difference between stabilizing CO_2 emissions and stabilizing atmospheric concentrations of CO_2. Stabilizing emissions of CO_2 at current levels would not lead to a stabilization in the atmospheric concentration of CO_2. In fact, stabilizing emissions at current levels would result in the atmospheric concentration of CO_2 continuing to rise over the 21st century and beyond.

The reason for this is that human activities are adding CO_2 to the atmosphere faster than natural processes can remove it. This is analogous to a flow of water into a bathtub. So long as the tap runs water (analogous to the emission of carbon dioxide) into the tub faster than water escapes through the plughole (analogous to the natural removal of carbon dioxide from the atmosphere) the level of water in the tub (analogous to the concentration of carbon dioxide in the atmosphere) will continue to rise.

According to some studies, stabilizing atmospheric CO_2 concentrations would require anthropogenic CO_2 emissions to be reduced by 80% relative to the peak emissions level. An 80% reduction in emissions would stabilize CO_2 concentrations for around a century, but even greater reductions would be required beyond this. Other research has found that, after leaving room for emissions for food production for 9 billion people and to keep the global temperature rise below 2 °C, emissions from energy production and transport will have to peak almost immediately in the developed world and decline at ca. 10% per annum until zero emissions are reached around 2030. In developing countries energy and transport emissions would have to peak by 2025 and then decline similarly. Stabilizing the atmospheric concentration of the other greenhouse gasses humans emit also depends on how fast their emissions are added to the atmosphere, and how fast the GHGs are removed.

In 2018, an international team of scientist published research saying that the current mitigation policy in Paris Agreement is insufficient to limit the temperature rise to 2 degrees. They say that even if all the current pledges will be accomplished there is a chance for a 4.5 degree temperature rise in decades. To preventing that, restoration of natural Carbon sinks, Carbon dioxide removal, changes in society and values will be necessary.

Projections

Projections of future greenhouse gas emissions are highly uncertain. In the absence of policies to mitigate climate change, GHG emissions could rise significantly over the 21st century.

Numerous assessments have considered how atmospheric GHG concentrations could be stabilized. The lower the desired stabilization level, the sooner global GHG emissions must peak and decline. GHG concentrations are unlikely to stabilize this century without major policy changes.

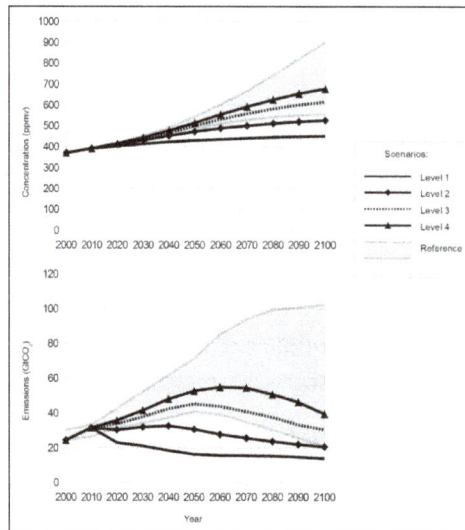

Projected carbon dioxide emissions and atmospheric concentrations over the 21st century for reference and mitigation scenarios.

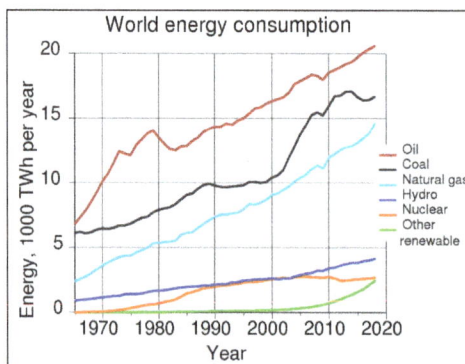

Rate of world energy usage per day, from 1970 to 2010.

Every fossil fuel source has increased in large amounts between 1970 and 2010, dominating all other energy sources. Hydroelectricity has increased at a slow steady rate over this same period, nuclear entered a period of rapid growth between 1970 and 1990 before leveling off. Other renewables, between 2000 and 2010 have, having started from a low usage rate, began to enter into a period of rapid growth. 1000 TWh=1 PWh.

Energy Consumption by Power Source

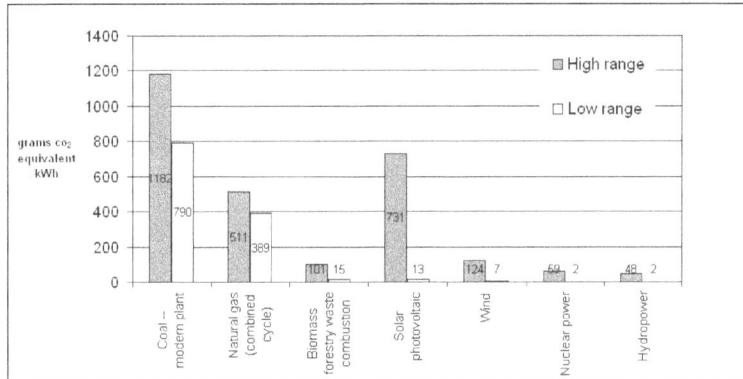

"Hydropower-Internalised Costs and Externalised Benefits"; Frans H. Koch; International Energy Agency (IEA)-Implementing Agreement for Hydropower Technologies and Programmes; 2000.

To create lasting climate change mitigation, the replacement of high carbon emission intensity power sources, such as conventional fossil fuels—oil, coal, and natural gas—with low-carbon power sources is required. Fossil fuels supply humanity with the vast majority of our energy demands, and at a growing rate. In 2012 the IEA noted that coal accounted for half the increased energy use of the prior decade, growing faster than all renewable energy sources. Both hydroelectricity and nuclear power together provide the majority of the generated low-carbon power fraction of global total power consumption.

Fuel type	Average total global power consumption in TW		
	1980	2004	2006
Oil	4.38	5.58	5.74
Gas	1.80	3.45	3.61
Coal	2.34	3.87	4.27
Hydroelectric	0.60	0.93	1.00
Nuclear power	0.25	0.91	0.93
Geothermal, wind, solar energy, wood	0.02	0.13	0.16
Total	9.48	15.0	15.8

Change and use of energy, by source, in units of (PWh) in that year.				
	Fossil	Nuclear	All renewables	Total
1990	83.374	6.113	13.082	102.569
2000	94.493	7.857	15.337	117.687
2008	117.076	8.283	18.492	143.851
Change 2000–2008	22.583	0.426	3.155	26.164

Methods and Means

Assessments often suggest that GHG emissions can be reduced using a portfolio of low-carbon technologies. At the core of most proposals is the reduction of greenhouse gas (GHG) emissions through reducing energy waste and switching to low-carbon power sources of energy. As the cost of reducing GHG emissions in the electricity sector appears to be lower than in other sectors, such as in the transportation sector, the electricity sector may deliver the largest proportional carbon reductions under an economically efficient climate policy.

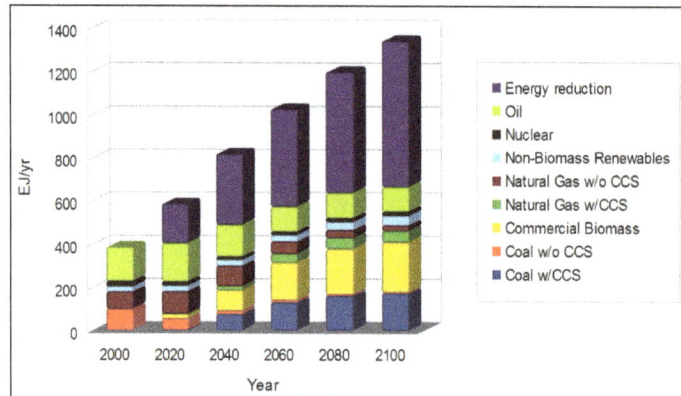

This graph shows the projected contribution of various energy sources to world primary electricity consumption (PEC). It is based on a climate change mitigation scenario, in which GHG emissions are substantially reduced over the 21st century. In the scenario, emission reductions are achieved using a portfolio of energy sources, as well as reductions in energy demand.

"Economic tools can be useful in designing climate change mitigation policies." "While the limitations of economics and social welfare analysis, including cost–benefit analysis, are widely documented, economics nevertheless provides useful tools for assessing the pros and cons of taking, or not taking, action on climate change mitigation, as well as of adaptation measures, in achieving competing societal goals. Understanding these pros and cons can help in making policy decisions on climate change mitigation and can influence the actions taken by countries, institutions and individuals."

Other frequently discussed means include efficiency, public transport, increasing fuel economy in automobiles (which includes the use of electric hybrids), charging plug-in hybrids and electric cars by low-carbon electricity, making individual changes, and changing business practices. Many fossil fuel driven vehicles can be converted to use electricity, the US has the potential to supply electricity for 73% of light duty vehicles (LDV), using overnight charging. The US average CO_2 emissions for a battery-electric car is 180 grams per mile vs 430 grams per mile for a gasoline car. The emissions would be displaced away from street level, where they have "high human-health implications. Increased use of electricity "generation for meeting the future transportation load is primarily fossil-fuel based", mostly natural gas, followed by coal, but could also be met through nuclear, tidal, hydroelectric and other sources.

A range of energy technologies may contribute to climate change mitigation. These include nuclear power and renewable energy sources such as biomass, hydroelectricity, wind power, solar power, geothermal power, ocean energy, and the use of carbon sinks, and carbon capture and storage. For example, Pacala and Socolow of Princeton have proposed a 15 part program to reduce CO_2

emissions by 1 billion metric tons per year – or 25 billion tons over the 50-year period using today's technologies as a type of global warming game.

Another consideration is how future socioeconomic development proceeds. Development choices (or "pathways") can lead differences in GHG emissions. Political and social attitudes may affect how easy or difficult it is to implement effective policies to reduce emissions.

Demand Side Management

Lifestyle and Behavior

The IPCC Fifth Assessment Report emphasises that behaviour, lifestyle, and cultural change have a high mitigation potential in some sectors, particularly when complementing technological and structural change. In general, higher consumption lifestyles have a greater environmental impact. Several scientific studies have shown that when people, especially those living in developed countries but more generally including all countries, wish to reduce their carbon footprint, there are four key "high-impact" actions they can take:

1. Not having an additional child (58.6 tonnes CO_2-equivalent emission reductions per year),

2. Living car-free (2.4 tonnes CO_2),

3. Avoiding one round-trip transatlantic flight (1.6 tonnes),

4. Eating a plant-based diet (0.8 tonnes).

These appear to differ significantly from the popular advice for "greening" one's lifestyle, which seem to fall mostly into the "low-impact" category: Replacing a typical car with a hybrid (0.52 tonnes); Washing clothes in cold water (0.25 tonnes); Recycling (0.21 tonnes); Upgrading light bulbs (0.10 tonnes); etc. The researchers found that public discourse on reducing one's carbon footprint overwhelmingly focuses on low-impact behaviors, and that mention of the high-impact behaviors is almost non-existent in the mainstream media, government publications, K-12 school textbooks, etc.

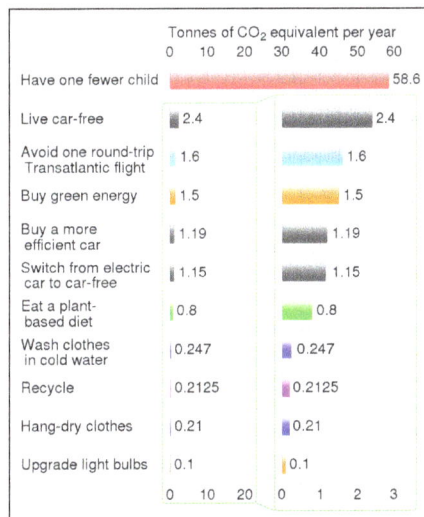

Reduction in one's carbon footprint for various actions.

The researchers added that "Our recommended high-impact actions are more effective than many more commonly discussed options (e.g. eating a plant-based diet saves eight times more emissions than upgrading light bulbs). More significantly, a US family who chooses to have one fewer child would provide the same level of emissions reductions as 684 teenagers who choose to adopt comprehensive recycling for the rest of their lives."

Dietary Change

Overall, food accounts for the largest share of consumption-based GHG emissions with nearly 20% of the global carbon footprint, followed by housing, mobility, services, manufactured products, and construction. Food and services are more significant in poor countries, while mobility and manufactured goods are more significant in rich countries. A 2014 study into the real-life diets of British people estimates their greenhouse gas contributions (CO_2eq) to be: 7.19 kg/day for high meat-eaters through to 3.81 kg/day for vegetarians and 2.89 kg/day for vegans. The widespread adoption of a vegetarian diet could cut food-related greenhouse gas emissions by 63% by 2050. China introduced new dietary guidelines in 2016 which aim to cut meat consumption by 50% and thereby reduce greenhouse gas emissions by 1 billion tonnes by 2030. A 2016 study concluded that taxes on meat and milk could simultaneously result in reduced greenhouse gas emissions and healthier diets. The study analyzed surcharges of 40% on beef and 20% on milk and suggests that an optimum plan would reduce emissions by 1 billion tonnes per year.

Energy Efficiency and Conservation

A 230-volt LED light bulb, with an E27 base (10 watts, 806 lumens).

Efficient energy use, sometimes simply called "energy efficiency", is the goal of efforts to reduce the amount of energy required to provide products and services. For example, insulating a home allows a building to use less heating and cooling energy to achieve and maintain a comfortable temperature. Installing LED lighting, fluorescent lighting, or natural skylight windows reduces the amount of energy required to attain the same level of illumination compared to using traditional incandescent light bulbs. Compact fluorescent lamps use only 33% of the energy and may last 6 to 10 times longer than incandescent lights. LED lamps use only about 10% of the energy an incandescent lamp requires.

Energy efficiency has proved to be a cost-effective strategy for building economies without necessarily growing energy consumption. For example, the state of California began implementing energy-efficiency measures in the mid-1970s, including building code and appliance standards with strict efficiency requirements. During the following years, California's energy consumption has remained approximately flat on a per capita basis while national US consumption doubled. As part of its strategy, California implemented a "loading order" for new energy resources that puts energy efficiency first, renewable electricity supplies second, and new fossil-fired power plants last.

Energy conservation is broader than energy efficiency in that it encompasses using less energy to achieve a lesser energy demanding service, for example through behavioral change, as well as encompassing energy efficiency. Examples of conservation without efficiency improvements would be heating a room less in winter, driving less, or working in a less brightly lit room. As with other definitions, the boundary between efficient energy use and energy conservation can be fuzzy, but both are important in environmental and economic terms. This is especially the case when actions are directed at the saving of fossil fuels.

Reducing energy use is seen as a key solution to the problem of reducing greenhouse gas emissions. According to the International Energy Agency, improved energy efficiency in buildings, industrial processes and transportation could reduce the world's energy needs in 2050 by one third, and help control global emissions of greenhouse gases.

Demand Side Switching Sources

Fuel switching on the demand side refers to changing the type of fuel used to satisfy a need for an energy service. To meet deep decarbonization goals, many primary energy changes are needed. Energy efficiency alone may not be sufficient to meet these goals, switching fuels used on the demand side will help lower carbon emissions. Progressively coal, oil and eventually natural gas for space and water heating in buildings will need to be reduced. For an equivalent amount of heat, burning natural gas produces about 45 per cent less carbon dioxide than burning coal. There are various ways in which this could happen, and different strategies will likely make sense in different locations. While the system efficiency of a gas furnace may be higher than the combination of natural gas power plant and electric heat, the combination of the same natural gas power plant and an electric heat pump has lower emissions per unit of heat delivered in all but the coldest climates. This is possible because of the very efficient coefficient of performance of heat pumps.

At the beginning of this century 70% of all electricity was generated by fossil fuels, and as carbon free sources eventually make up half of the generation mix, replacing gas or oil furnaces and water heaters with electric ones will have a climate benefit. In areas like Norway, Brazil, and Quebec that have abundant hydroelectricity, electric heat and hot water are common.

The economics of switching the demand side from fossil fuels to electricity for heating, will depend on the price of fuels vs electricity and the relative prices of the equipment. The EIA Annual Energy Outlook 2014 suggests that domestic gas prices will rise faster than electricity prices which will encourage electrification in the coming decades. Electrifying heating loads may also provide a flexible resource that can participate in demand response. Since, thermostatically controlled loads have inherent energy storage, electrification of heating could provide a valuable resource to integrate variable renewable resources into the grid.

Alternatives to electrification, include decarbonizing pipeline gas through power to gas, biogas, or other carbon-neutral fuels. A 2015 study by Energy+Environmental Economics shows that a hybrid approach of decarbonizing pipeline gas, electrification, and energy efficiency can meet carbon reduction goals at a similar cost as only electrification and energy efficiency in Southern California.

Demand Side Grid Management

Expanding intermittent electrical sources such as wind power, creates a growing problem balancing grid fluctuations. Some of the plans include building pumped storage or continental super grids costing billions of dollars. However instead of building for more power, there are a variety of ways to affect the size and timing of electricity demand on the consumer side. Designing for reduced demands on a smaller power grid is more efficient and economic than having extra generation and transmission for intermittentcy, power failures and peak demands. Having these abilities is one of the chief aims of a smart grid.

Time of use metering is a common way to motivate electricity users to reduce their peak load consumption. For instance, running dishwashers and laundry at night after the peak has passed, reduces electricity costs.

Dynamic demand plans have devices passively shut off when stress is sensed on the electrical grid. This method may work very well with thermostats, when power on the grid sags a small amount, a low power temperature setting is automatically selected reducing the load on the grid. For instance millions of refrigerators reduce their consumption when clouds pass over solar installations. Consumers would need to have a smart meter in order for the utility to calculate credits.

Demand response devices could receive all sorts of messages from the grid. The message could be a request to use a low power mode similar to dynamic demand, to shut off entirely during a sudden failure on the grid, or notifications about the current and expected prices for power. This would allow electric cars to recharge at the least expensive rates independent of the time of day. The vehicle-to-grid suggestion would use a car's battery or fuel cell to supply the grid temporarily.

Alternative Energy Sources

Renewable Energy

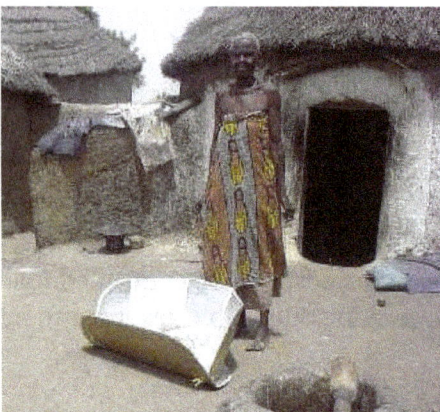

Solar cookers use sunlight as energy source for outdoor cooking.

The 22,500 MW nameplate capacity Three Gorges Dam in the People's Republic of China, the largest hydroelectric power station in the world.

Renewable energy flows involve natural phenomena such as sunlight, wind, rain, tides, plant growth, and geothermal heat, as the International Energy Agency explains:

> Renewable energy is derived from natural processes that are replenished constantly. In its various forms, it derives directly from the sun, or from heat generated deep within the earth. Included in the definition is electricity and heat generated from solar, wind, ocean, hydropower, biomass, geothermal resources, and biofuels and hydrogen derived from renewable resources.

Climate change concerns and the need to reduce carbon emissions are driving increasing growth in the renewable energy industries. Low-carbon renewable energy replaces conventional fossil fuels in three main areas: power generation, hot water/space heating, and transport fuels. In 2011, the share of renewables in electricity generation worldwide grew for the fourth year in a row to 20.2%. Based on REN21's 2014 report, renewables contributed 19% to supply global energy consumption. This energy consumption is divided as 9% coming from burning biomass, 4.2% as heat energy (non-biomass), 3.8% hydro electricity and 2% as electricity from wind, solar, geothermal, and biomass thermal power plants.

Renewable energy use has grown much faster than anyone anticipated. The Intergovernmental Panel on Climate Change (IPCC) has said that there are few fundamental technological limits to integrating a portfolio of renewable energy technologies to meet most of total global energy demand. At the national level, at least 30 nations around the world already have renewable energy contributing more than 20% of energy supply.

As of 2012, renewable energy accounts for almost half of new electricity capacity installed and costs are continuing to fall. Public policy and political leadership helps to "level the playing field" and drive the wider acceptance of renewable energy technologies. As of 2011, 118 countries have targets for their own renewable energy futures, and have enacted wide-ranging public policies to promote renewables. Leading renewable energy companies include BrightSource Energy, First Solar, Gamesa, GE Energy, Goldwind, Sinovel, Suntech, Trina Solar, Vestas, and Yingli.

The incentive to use 100% renewable energy has been created by global warming and other ecological as well as economic concerns. Mark Z. Jacobson says producing all new energy with wind power, solar power, and hydropower by 2030 is feasible and existing energy supply arrangements could be replaced by 2050. Barriers to implementing the renewable energy plan are seen to be "primarily social and political, not technological or economic". Jacobson says that energy costs with a wind, solar, water system should be similar to today's energy costs. According to a 2011 projection by the (IEA) International Energy Agency, solar power generators may produce most of the world's electricity within 50 years, dramatically reducing harmful greenhouse gas emissions. Critics of the "100% renewable energy" approach include Vaclav Smil and James E. Hansen. Smil and Hansen are concerned about the variable output of solar and wind power, NIMBYism, and a lack of infrastructure.

Economic analysts expect market gains for renewable energy (and efficient energy use) following the 2011 Japanese nuclear accidents. In his 2012 State of the Union address, President Barack Obama restated his commitment to renewable energy and mentioned the long-standing Interior Department commitment to permit 10,000 MW of renewable energy projects on public land in

2012. Globally, there are an estimated 3 million direct jobs in renewable energy industries, with about half of them in the biofuels industry.

Some countries, with favorable geography, geology, and weather well suited to an economical exploitation of renewable energy sources, already get most of their electricity from renewables, including from geothermal energy in Iceland (100 percent), and hydroelectric power in Brazil (85 percent), Austria (62 percent), New Zealand (65 percent), and Sweden (54 percent). Renewable power generators are spread across many countries, with wind power providing a significant share of electricity in some regional areas: for example, 14 percent in the US state of Iowa, 40 percent in the northern German state of Schleswig-Holstein, and 20 percent in Denmark. Solar water heating makes an important and growing contribution in many countries, most notably in China, which now has 70 percent of the global total (180 GWth). Worldwide, total installed solar water heating systems meet a portion of the water heating needs of over 70 million households. The use of biomass for heating continues to grow as well. In Sweden, national use of biomass energy has surpassed that of oil. Direct geothermal heating is also growing rapidly.

Renewable biofuels for transportation, such as ethanol fuel and biodiesel, have contributed to a significant decline in oil consumption in the United States since 2006. The 93 billion liters of biofuels produced worldwide in 2009 displaced the equivalent of an estimated 68 billion liters of gasoline, equal to about 5 percent of world gasoline production. Many different biofuel generations can be distinguished, namely 1st, 2nd, 3rd and 4th generation biofuel. Whereas first generation biofuels competed with food production, the later generations no longer had that problem. Also, as 1st generation biofuels also comprise such fuels as palm oil and soy oil (which are drivers for deforestation in rainforests (Brazil and Indonesia) that issue is also no longer present in later-generation biofuels.

Some of the world's largest solar power stations: Ivanpah (CSP) and Topaz (PV), both in California.

Nuclear Power

Since about 2001 the term "nuclear renaissance" has been used to refer to a possible nuclear power industry revival, driven by rising fossil fuel prices and new concerns about meeting greenhouse gas emission limits. However, in March 2011 the Fukushima nuclear disaster in Japan and associated shutdowns at other nuclear facilities raised questions among some commentators over the future of nuclear power. Platts has reported that "the crisis at Japan's Fukushima nuclear plants has prompted leading energy-consuming countries to review the safety of their existing reactors and cast doubt on the speed and scale of planned expansions around the world".

Blue Cherenkov light being produced near the core of the Fission powered Advanced Test Reactor.

The World Nuclear Association has reported that nuclear electricity generation in 2012 was at its lowest level since 1999. Several previous international studies and assessments, suggested that as part of the portfolio of other low-carbon energy technologies, nuclear power will continue to play a role in reducing greenhouse gas emissions. Historically, nuclear power usage is estimated to have prevented the atmospheric emission of 64 gigatonnes of CO_2-equivalent as of 2013. Public concerns about nuclear power include the fate of spent nuclear fuel, nuclear accidents, security risks, nuclear proliferation, and a concern that nuclear power plants are very expensive. Of these concerns, nuclear accidents and disposal of long-lived radioactive fuel/waste have probably had the greatest public impact worldwide. Although generally unaware of it, both of these glaring public concerns are greatly diminished by present passive safety designs, the experimentally proven, "melt-down proof" EBR-II, future molten salt reactors, and the use of conventional and more advanced fuel/waste pyroprocessing, with the latter recycling or reprocessing not presently being commonplace as it is often considered to be cheaper to use a once-through nuclear fuel cycle in many countries, depending on the varying levels of intrinsic value given by a society in reducing the long-lived waste in their country, with France doing a considerable amount of reprocessing when compared to the US.

Nuclear power, with a 10.6% share of world electricity production as of 2013, is second only to hydroelectricity as the largest source of low-carbon power. Over 400 reactors generate electricity in 31 countries.

While some have raised uncertainty surrounding the future GHG emissions of nuclear power as a result of an extreme potential decline in uranium ore grade without a corresponding increase in the efficiency of enrichment methods. In a scenario analysis of future global nuclear development, as it could be effected by a decreasing global uranium market of average ore grade, the analysis determined that depending on conditions, median life cycle nuclear power GHG emissions could be between 9 and 110 g CO_2-eq/kWh by 2050, with the latter high figure being derived from a "worst-case scenario" that is not "considered very robust" by the authors of the paper, as the "ore grade" in the scenario is lower than the uranium concentration in many lignite coal ashes.

Although this future analyses primarily deals with extrapolations for present Generation II reactor technology, of which two are in operation as of 2014 with the newest being the BN-800, for these reactors it states that the "median life cycle GHG emissions are similar to or lower than [present light water reactors] LWRs and purports to consume little or no uranium ore.

In their 2014 report, the IPCC comparison of energy sources global warming potential per unit of electricity generated, which notably included albedo effects, mirror the median emission value derived from the Warner and Heath Yale meta-analysis for the more common non-breeding light water reactors, a CO_2-equivalent value of 12 g CO_2-eq/kWh, which is the lowest global warming forcing of all baseload power sources, with comparable low carbon power baseload sources, such as hydropower and biomass, producing substantially more global warming forcing 24 and 230 g CO_2-eq/kWh respectively.

In 2014, Brookings Institution published The Net Benefits of Low and No-Carbon Electricity Technologies which states, after performing an energy and emissions cost analysis, that "The net benefits of new nuclear, hydro, and natural gas combined cycle plants far outweigh the net benefits of new wind or solar plants", with the most cost effective low carbon power technology being determined to be nuclear power.

During his presidential campaign, Barack Obama stated, "Nuclear power represents more than 70% of our noncarbon generated electricity. It is unlikely that we can meet our aggressive climate goals if we eliminate nuclear power as an option."

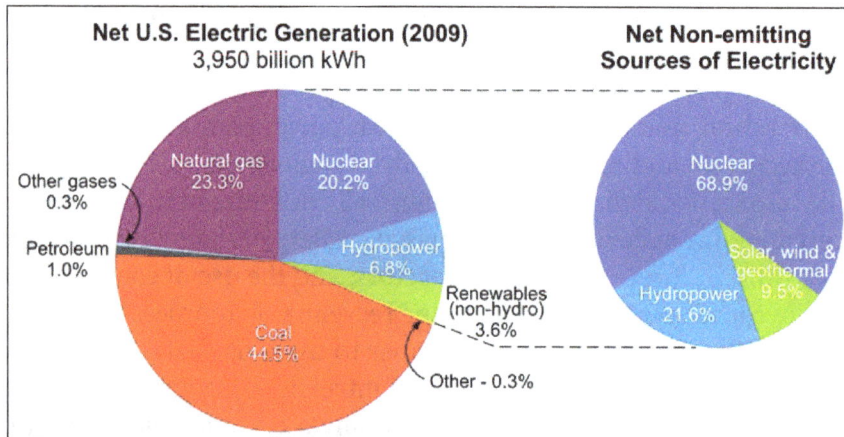

This graph illustrates nuclear power is the United States's largest contributor of non-greenhouse gas-emitting electric power generation, comprising nearly three-quarters of the non-emitting sources.

Analysis in 2015 by Professor and Chair of Environmental Sustainability Barry W. Brook and his colleagues on the topic of replacing fossil fuels entirely, from the electric grid of the world, has determined that at the historically modest and proven-rate at which nuclear energy was added to and replaced fossil fuels in France and Sweden during each nation's building programs in the 1980s, within 10 years nuclear energy could displace or remove fossil fuels from the electric grid completely, "allowing the world to meet the most stringent greenhouse-gas mitigation targets." In a similar analysis, Brook had earlier determined that 50% of all global energy, that is not solely electricity, but transportation synfuels etc. could be generated within approximately 30 years, if the global nuclear fission build rate was identical to each of these nation's already proven decadal rates (in units of installed nameplate capacity, GW per year, per unit of global GDP (GW/year/$).

This is in contrast to the completely conceptual paper-studies for a 100% renewable energy world, which would require an orders of magnitude more costly global investment per year, an investment rate that has no historical precedent, having never been attempted due to its prohibitive cost, and with far greater land area that would be required to be devoted to the wind, wave and solar projects, along with the inherent assumption that humanity will use less, and not more, energy in the future. As Brook notes the "principal limitations on nuclear fission are not technical, economic or fuel-related, but are instead linked to complex issues of societal acceptance, fiscal and political inertia, and inadequate critical evaluation of the real-world constraints facing (the other) low-carbon alternatives."

Nuclear power may be uncompetitive compared with fossil fuel energy sources in countries without a carbon tax program, and in comparison to a fossil fuel plant of the same power output, nuclear power plants take a longer amount of time to construct.

Two new, first of their kind, EPR reactors under construction in Finland and France have been delayed and are running over-budget. However learning from experience, two further EPR reactors under construction in China are on, and ahead, of schedule respectively. As of 2013, according to the IAEA and the European Nuclear Society, worldwide there were 68 civil nuclear power reactors under construction in 15 countries. China has 29 of these nuclear power reactors under construction, as of 2013, with plans to build many more, while in the US the licenses of almost half its reactors have been extended to 60 years, and plans to build another dozen are under serious consideration. There are also a considerable number of new reactors being built in South Korea, India, and Russia. At least 100 older and smaller reactors will "most probably be closed over the next 10–15 years". This is probable only if one does not factor in the ongoing Light Water Reactor Sustainability Program, created to permit the extension of the life span of the USA's 104 nuclear reactors to 60 years. The licenses of almost half of the USA's reactors have been extended to 60 years as of 2008. Two new "passive safety" AP1000 reactors are, as of 2013, being constructed at Vogtle Electric Generating Plant.

Public opinion about nuclear power varies widely between countries. A poll by Gallup International assessed public opinion in 47 countries. The poll was conducted following a tsunami and earthquake which caused an accident at the Fukushima nuclear power plant in Japan. 49% stated that they held favourable views about nuclear energy, while 43% held an unfavourable view. Another global survey by Ipsos assessed public opinion in 24 countries. Respondents to this survey showed a clear preference for renewable energy sources over coal and nuclear energy. Ipsos found that solar and wind were viewed by the public as being more environmentally friendly and more viable long-term energy sources relative to nuclear power and natural gas. However, solar and wind were viewed as being less reliable relative to nuclear power and natural gas. In 2012 a poll done in the UK found that 63% of those surveyed support nuclear power, and with opposition to nuclear power at 11%. In Germany, strong anti-nuclear sentiment led to eight of the seventeen operating reactors being permanently shut down following the March 2011 Fukushima nuclear disaster.

Nuclear fusion research, in the form of the International Thermonuclear Experimental Reactor is underway. Fusion powered electricity generation was initially believed to be readily achievable, as fission power had been. However, the extreme requirements for continuous reactions and plasma containment led to projections being extended by several decades. In 2010, more than 60 years after the first attempts, commercial power production was still believed to be unlikely before 2050. Although rather than an either, or, issue economical fusion-fission hybrid reactors could be built before any attempt at this more demanding commercial pure-fusion reactor/DEMO reactor takes place.

Coal to Gas Fuel Switching

Most mitigation proposals imply—rather than directly state—an eventual reduction in global fossil fuel production. Also proposed are direct quotas on global fossil fuel production.

Natural gas emits far fewer greenhouse gases (i.e. CO_2 and methane—CH_4) than coal when burned at power plants, but evidence has been emerging that this benefit could be completely negated by methane leakage at gas drilling fields and other points in the supply chain.

A study performed by the Environmental Protection Agency (EPA) and the Gas Research Institute (GRI) in 1997 sought to discover whether the reduction in carbon dioxide emissions from increased natural gas (predominantly methane) use would be offset by a possible increased level of methane emissions from sources such as leaks and emissions. The study concluded that the reduction in emissions from increased natural gas use outweighs the detrimental effects of increased methane emissions. More recent peer-reviewed studies have challenged the findings of this study, with researchers from the National Oceanic and Atmospheric Administration (NOAA) reconfirming findings of high rates of methane (CH_4) leakage from natural gas fields.

A 2011 study by noted climate research scientist, Tom Wigley, found that while carbon dioxide (CO_2) emissions from fossil fuel combustion may be reduced by using natural gas rather than coal to produce energy, it also found that additional methane (CH_4) from leakage adds to the radiative forcing of the climate system, offsetting the reduction in CO_2 forcing that accompanies the transition from coal to gas. The study looked at methane leakage from coal mining; changes in radiative forcing due to changes in the emissions of sulfur dioxide and carbonaceous aerosols; and differences in the efficiency of electricity production between coal- and gas-fired power generation. On balance, these factors more than offset the reduction in warming due to reduced CO_2 emissions. When gas replaces coal there is additional warming out to 2,050 with an assumed leakage rate of 0%, and out to 2,140 if the leakage rate is as high as 10%. The overall effects on global-mean temperature over the 21st century, however, are small. Petron et al. and Alvarez et al. note that estimated that leakage from gas infrastructure is likely to be underestimated. These studies indicate that the exploitation of natural gas as a "cleaner" fuel is questionable. A 2014 meta-study of 20 years of natural gas technical literature shows that methane emissions are consistently underestimated but on a 100-year scale, the climate benefits of coal to gas fuel switching are likely larger than the negative effects of natural gas leakage.

Heat Pump

A heat pump is a device that provides heat energy from a source of heat to a destination called a "heat sink". Heat pumps are designed to move thermal energy opposite to the direction of spontaneous heat flow by absorbing heat from a cold space and releasing it to a warmer one. A heat pump uses some amount of external power to accomplish the work of transferring energy from the heat source to the heat sink.

While air conditioners and freezers are familiar examples of heat pumps, the term "heat pump" is more general and applies to many HVAC (heating, ventilating, and air conditioning) devices used for space heating or space cooling. When a heat pump is used for heating, it employs the same basic refrigeration-type cycle used by an air conditioner or a refrigerator, but in the opposite direction—releasing heat into the conditioned space rather than the surrounding environment. In this use, heat pumps generally draw heat from the cooler external air or from the ground. In heating mode, heat pumps are three to four times more efficient in their use of electric power than simple electrical resistance heaters.

Outside unit of an air-source heat pump.

It has been concluded that heat pumps are the single technology that could reduce the greenhouse gas emissions of households better than every other technology that is available on the market. With a market share of 30% and (potentially) clean electricity, heat pumps could reduce global CO_2 emissions by 8% annually. Using ground source heat pumps could reduce around 60% of the primary energy demand and 90% of CO_2 emissions in Europe in 2050 and make handling high shares of renewable energy easier. Using surplus renewable energy in heat pumps is regarded as the most effective household means to reduce global warming and fossil fuel depletion.

With significant amounts of fossil fuel used in electricity production, demands on the electrical grid also generate greenhouse gases. Without a high share of low-carbon electricity, a domestic heat pump will produce more carbon emissions than using natural gas.

Fossil Fuel Phase-out: Carbon Neutral and Negative Fuels

3,500–4,000 environmental activists blocking a coal mine
in Germany to limit climate change.

Fossil fuel may be phased-out with carbon-neutral and carbon-negative pipeline and transportation fuels created with power to gas and gas to liquids technologies. Carbon dioxide from fossil fuel flue gas can be used to produce plastic lumber allowing carbon negative reforestation.

Sinks and Negative Emissions

A carbon sink is a natural or artificial reservoir that accumulates and stores some carbon-containing chemical compound for an indefinite period, such as a growing forest. A negative carbon dioxide emission on the other hand is a permanent removal of carbon dioxide out of the atmosphere. Examples are direct air capture, enhanced weathering technologies such as storing it in geologic formations underground and biochar. These processes are sometimes considered as variations of sinks or mitigation, and sometimes as geoengineering. In combination with other mitigation measures, sinks in combination with negative carbon emissions are considered crucial for meeting the 350 ppm target.

The Antarctic Climate and Ecosystems Cooperative Research Centre (ACE-CRC) notes that one third of humankind's annual emissions of CO_2 are absorbed by the oceans. However, this also leads to ocean acidification, with potentially significant impacts on marine life. Acidification lowers the level of carbonate ions available for calcifying organisms to form their shells. These organisms include plankton species that contribute to the foundation of the Southern Ocean food web. However acidification may impact on a broad range of other physiological and ecological processes, such as fish respiration, larval development and changes in the solubility of both nutrients and toxins. Some plants such as seaweed generate oxygen, are farmed, and can act as a source of (third-generation) biofuel (hereby temporarily sequestering carbon as well).

Reforestation, Avoided Deforestation and Afforestation

According to a research by Tom Crowther et al, there is still enough room to plant an additional 1.2 trillion trees. This amount of trees would cancel out the last 10 years of CO_2 emissions and sequester 160 billion tons of carbon. This vision is being executed by the Trillion Tree Campaign. According to research conducted at ETH Zurich, restoring all degraded forests all over the world could capture about 205 billion tons of carbon in total (which is about 2/3rd of all carbon emissions, bringing global warming down to below 2 °C).

Transferring land rights to indigenous inhabitants is argued to efficiently conserve forests.

Almost 20 percent (8 $GtCO_2$/year) of total greenhouse-gas emissions were from deforestation in 2007. It is estimated that avoided deforestation reduces CO_2 emissions at a rate of 1 tonne of CO_2 per $1–5 in opportunity costs from lost agriculture. Reforestation could save at least another 1 $GtCO_2$/year, at an estimated cost of $5–15/$tCO_2$. Afforestation is where there was previously no forest – such plantations are estimated to have to be prohibitively massive to be reduce emissions by itself.

Transferring rights over land from public domain to its indigenous inhabitants, who have had a stake for millennia in preserving the forests that they depend on, is argued to be a cost effective strategy to conserve forests. This includes the protection of such rights entitled in existing laws, such as India's Forest Rights Act. The transferring of such rights in China, perhaps the largest land reform in modern times, has been argued to have increased forest cover. Granting title of the land has shown to have two or three times less clearing than even state run parks, notably in the Brazilian Amazon. Excluding humans and even evicting inhabitants from protected areas (called "fortress conservation"), sometimes as a result of lobbying by environmental groups, often lead to more exploitation of the land as the native inhabitants then turn to work for extractive companies to survive.

With increased intensive agriculture and urbanization, there is an increase in the amount of abandoned farmland. By some estimates, for every half a hectare of original old-growth forest cut down, more than 20 hectares of new secondary forests are growing, even though they do not have the same biodiversity as the original forests and original forests store 60% more carbon than these new secondary forests. According to a study, promoting regrowth on abandoned farmland could offset years of carbon emissions. Research by the university ETH Zurich estimates that Russia, the United States and Canada have the most land suitable for reforestation.

Avoided Desertification

Restoring grasslands store CO_2 from the air into plant material. Grazing livestock, usually not left to wander, would eat the grass and would minimize any grass growth. However, grass left alone would eventually grow to cover its own growing buds, preventing them from photosynthesizing and the dying plant would stay in place. A method proposed to restore grasslands uses fences with many small paddocks and moving herds from one paddock to another after a day a two in order to mimic natural grazers and allowing the grass to grow optimally. Additionally, when part of leaf matter is consumed by a herding animal, a corresponding amount of root matter is sloughed off too as it would not be able to sustain the previous amount of root matter and while most of the lost root matter would rot and enter the atmosphere, part of the carbon is sequestered into the soil. It is estimated that increasing the carbon content of the soils in the world's 3.5 billion hectares of agricultural grassland by 1% would offset nearly 12 years of CO_2 emissions. Allan Savory, as part of holistic management, claims that while large herds are often blamed for desertification, prehistoric lands supported large or larger herds and areas where herds were removed in the United States are still desertifying.

Managed grazing methods are argued to be able to restore grasslands, thereby significantly decreasing atmospheric CO_2 levels.

Additionally, the global warming induced thawing of the permafrost, which stores about two times the amount of the carbon currently released in the atmosphere, releases the potent greenhouse gas, methane, in a positive feedback cycle that is feared to lead to a tipping point called runaway climate change. A method proposed to prevent such a scenario is to bring back large herbivores such as seen in Pleistocene Park, where their trampling naturally keep the ground cooler by eliminating shrubs and keeping the ground exposed to the cold air.

Carbon Capture and Storage

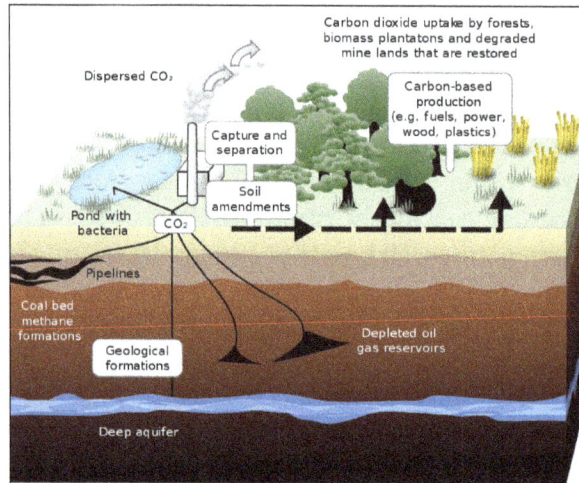

Schematic showing both terrestrial and geological sequestration of carbon dioxide emissions from a coal-fired plant.

Carbon capture and storage (CCS) is a method to mitigate climate change by capturing carbon dioxide (CO_2) from large point sources such as power plants and subsequently storing it away safely instead of releasing it into the atmosphere. The IPCC estimates that the costs of halting global warming would double without CCS. The International Energy Agency says CCS is "the most important single new technology for CO_2 savings" in power generation and industry. Though it requires up to 40% more energy to run a CCS coal power plant than a regular coal plant, CCS could potentially capture about 90% of all the carbon emitted by the plant. Norway's Sleipner gas field, beginning in 1996, stores almost a million tons of CO_2 a year to avoid penalties in producing natural gas with unusually high levels of CO_2. As of late 2011, the total planned CO_2 storage capacity of all 14 projects in operation or under construction is over 33 million tonnes a year. This is broadly equivalent to preventing the emissions from more than six million cars from entering the atmosphere each year. According to a Sierra Club analysis, the US coal fired Kemper Project due to be online in 2017, is the most expensive power plant ever built for the watts of electricity it will generate.

Enhanced Weathering

Enhanced weathering is the removal of carbon from the air into the earth, enhancing the natural carbon cycle where carbon is mineralized into rock. The CarbFix project couples with carbon capture and storage in power plants to turn carbon dioxide into stone in a relatively short period of two years, addressing the common concern of leakage in CCS projects. While this project used basalt rocks, olivine has also shown promise.

Geoengineering

Geoengineering is seen by Olivier Sterck as an alternative to mitigation and adaptation, but by Gernot Wagner as an entirely separate response to climate change. In a literature assessment, Barker *et al.* described geoengineering as a type of mitigation policy. IPCC concluded that geoengineering options, such as ocean fertilization to remove CO_2 from the atmosphere, remained largely unproven. It was judged that reliable cost estimates for geoengineering had not yet been published.

The National Academy of Sciences report Policy Implications of Greenhouse Warming: Mitigation, Adaptation, and the Science Base defined geoengineering as "options that would involve large-scale engineering of our environment in order to combat or counteract the effects of changes in atmospheric chemistry." They evaluated a range of options to try to give preliminary answers to two questions: can these options work and could they be carried out with a reasonable cost. They also sought to encourage discussion of a third question — what adverse side effects might there be. The following types of option were examined: reforestation, increasing ocean absorption of carbon dioxide (carbon sequestration) and screening out some sunlight. NAS also argued "Engineered countermeasures need to be evaluated but should not be implemented without broad understanding of the direct effects and the potential side effects, the ethical issues, and the risks." In July 2011 a report by the United States Government Accountability Office on geoengineering found that "climate engineering technologies do not now offer a viable response to global climate change."

Carbon Dioxide Removal

Carbon dioxide removal has been proposed as a method of reducing the amount of radiative forcing. A variety of means of artificially capturing and storing carbon, as well as of enhancing natural sequestration processes, are being explored. The main natural process is photosynthesis by plants and single-celled organisms. Artificial processes vary, and concerns have been expressed about the long-term effects of some of these processes.

It is notable that the availability of cheap energy and appropriate sites for geological storage of carbon may make carbon dioxide air capture viable commercially. It is, however, generally expected that carbon dioxide air capture may be uneconomic when compared to carbon capture and storage from major sources — in particular, fossil fuel powered power stations, refineries, etc. As in the case of the US Kemper Project with carbon capture, costs of energy produced will grow significantly. However, captured CO_2 can be used to force more crude oil out of oil fields, as Statoil and Shell have made plans to do. CO_2 can also be used in commercial greenhouses, giving an opportunity to kick-start the technology. Some attempts have been made to use algae to capture smokestack emissions, notably the GreenFuel Technologies Corporation, who have now shut down operations.

Solar Radiation Management

The main purpose of solar radiation management seek to reflect sunlight and thus reduce global warming. The ability of stratospheric sulfate aerosols to create a global dimming effect has made them a possible candidate for use in climate engineering projects.

Non-CO$_2$ Greenhouse Gases

CO$_2$ is not the only GHG relevant to mitigation, and governments have acted to regulate the emissions of other GHGs emitted by human activities (anthropogenic GHGs). The emissions caps agreed to by most developed countries under the Kyoto Protocol regulate the emissions of almost all the anthropogenic GHGs. These gases are CO$_2$, methane (CH$_4$), nitrous oxide (N$_2$O), the hydrofluorocarbons (HFC), perfluorocarbons (PFC), and sulfur hexafluoride (SF$_6$).

Stabilizing the atmospheric concentrations of the different anthropogenic GHGs requires an understanding of their different physical properties. Stabilization depends both on how quickly GHGs are added to the atmosphere and how fast they are removed. The rate of removal is measured by the atmospheric lifetime of the GHG in question. Here, the lifetime is defined as the time required for a given perturbation of the GHG in the atmosphere to be reduced to 37% of its initial amount. Methane has a relatively short atmospheric lifetime of about 12 years, while N$_2$O's lifetime is about 110 years. For methane, a reduction of about 30% below current emission levels would lead to a stabilization in its atmospheric concentration, while for N$_2$O, an emissions reduction of more than 50% would be required.

Methane is a significantly more potent greenhouse gas than carbon dioxide in the amount of heat it can trap, especially in the short term. Burning one molecule of methane generates one molecule of carbon dioxide, indicating there may be no net benefit in using gas as a fuel source. Reducing the amount of waste methane produced in the first place and moving away from use of gas as a fuel source will have a greater beneficial impact, as might other approaches to productive use of otherwise-wasted methane. In terms of prevention, vaccines are being developed in Australia to reduce the significant global warming contributions from methane released by livestock via flatulence and eructation.

Another physical property of the anthropogenic GHGs relevant to mitigation is the different abilities of the gases to trap heat (in the form of infrared radiation). Some gases are more effective at trapping heat than others, e.g., SF$_6$ is 22,200 times more effective a GHG than CO$_2$ on a per-kilogram basis. A measure for this physical property is the global warming potential (GWP), and is used in the Kyoto Protocol.

Although not designed for this purpose, the Montreal Protocol has probably benefited climate change mitigation efforts. The Montreal Protocol is an international treaty that has successfully reduced emissions of ozone-depleting substances (for example, CFCs), which are also greenhouse gases.

By Sector

The Tesla Roadster emits no tailpipe emissions, uses lithium ion batteries to
achieve 220 mi (350 km) per charge, while also capable of going 0–60 in under 4 seconds.

Bicycles have almost no carbon footprint compared to cars, and canal transport may represent a positive option for certain types of freight in the 21st century.

Transport

Transportation emissions account for roughly 1/4 of emissions worldwide, and are even more important in terms of impact in developed nations especially in North America and Australia. Many citizens of countries like the United States and Canada who drive personal cars often, see well over half of their climate change impact stemming from the emissions produced from their cars. Modes of mass transportation such as bus, light rail (metro, subway, etc.), and long-distance rail are far and away the most energy-efficient means of motorized transportation for passengers, able to use in many cases over twenty times less energy per person-distance than a personal automobile. Modern energy-efficient technologies, such as plug-in hybrid electric vehicles and carbon-neutral synthetic gasoline & Jet fuel may also help to reduce the consumption of petroleum, land use changes and emissions of carbon dioxide. Utilizing rail transport, especially electric rail, over the far less efficient air transport and truck transport significantly reduces emissions. With the use of electric trains and cars in transportation there is the opportunity to run them with low-carbon power, producing far fewer emissions.

Urban Planning

Effective urban planning to reduce sprawl aims to decrease Vehicle Miles Travelled (VMT), lowering emissions from transportation. Personal cars are extremely inefficient at moving passengers, while public transport and bicycles are many times more efficient (as is the simplest form of human transportation, walking). All of these are encouraged by urban/community planning and are an effective way to reduce greenhouse gas emissions. Between 1982 and 1997, the amount of land consumed for urban development in the United States increased by 47 percent while the nation's population grew by only 17 percent. Inefficient land use development practices have increased infrastructure costs as well as the amount of energy needed for transportation, community services, and buildings.

At the same time, a growing number of citizens and government officials have begun advocating a smarter approach to land use planning. These smart growth practices include compact community development, multiple transportation choices, mixed land uses, and practices to conserve green space. These programs offer environmental, economic, and quality-of-life benefits; and they also serve to reduce energy usage and greenhouse gas emissions.

Approaches such as New Urbanism and transit-oriented development seek to reduce distances travelled, especially by private vehicles, encourage public transit and make walking and cycling more attractive options. This is achieved through "medium-density", mixed-use planning and the concentration of housing within walking distance of town centers and transport nodes.

Smarter growth land use policies have both a direct and indirect effect on energy consuming behavior. For example, transportation energy usage, the number one user of petroleum fuels, could be significantly reduced through more compact and mixed use land development patterns, which in turn could be served by a greater variety of non-automotive based transportation choices.

Building Design

Emissions from housing are substantial, and government-supported energy efficiency programmes can make a difference.

For institutions of higher learning in the United States, greenhouse gas emissions depend primarily on total area of buildings and secondarily on climate. If climate is not taken into account, annual greenhouse gas emissions due to energy consumed on campuses plus purchased electricity can be estimated with the formula, $E=aS^b$, where a =0.001621 metric tonnes of CO_2 equivalent/square foot or 0.0241 metric tonnes of CO_2 equivalent/square meter and b= 1.1354.

New buildings can be constructed using passive solar building design, low-energy building, or zero-energy building techniques, using renewable heat sources. Existing buildings can be made more efficient through the use of insulation, high-efficiency appliances (particularly hot water heaters and furnaces), double- or triple-glazed gas-filled windows, external window shades, and building orientation and siting. Renewable heat sources such as shallow geothermal and passive solar energy reduce the amount of greenhouse gasses emitted. In addition to designing buildings which are more energy-efficient to heat, it is possible to design buildings that are more energy-efficient to cool by using lighter-coloured, more reflective materials in the development of urban areas (e.g. by painting roofs white) and planting trees. This saves energy because it cools buildings and reduces the urban heat island effect thus reducing the use of air conditioning.

Agriculture

In the United States, soils account for about half of agricultural greenhouse gas emissions while agriculture, forestry and other land use emits 24%. Globally, livestock is responsible for 18 percent of greenhouse gas emissions, according to FAO's report called "Livestock's Long Shadow: Environmental Issues and Options".

The US EPA says soil management practices that can reduce the emissions of nitrous oxide (N_2O) from soils include fertilizer usage, irrigation, and tillage. Manure management and rice cultivation also produce gaseous emissions.

Important mitigation options for reducing the greenhouse gas emissions from livestock (especially ruminants) are genetic selection immunization, rumen defaunation, outcompetition of methanogenic archaea with acetogens, introduction of methanotrophic bacteria into the rumen, diet modification and grazing management, among others. Certain diet changes (ie. with Asparagopsis taxiformis) allow for a reduction of upto 99% of ruminant ghg emissions.

Other options include just using ruminant-free alternatives instead, such as milk substitutes and meat analogues.

Methods that enhance carbon sequestration in soil include no-till farming, residue mulching, cover cropping, and crop rotation, all of which are more widely used in organic farming than in conventional farming. Because only 5% of US farmland currently uses no-till and residue mulching, there is a large potential for carbon sequestration.

A 2015 study found that farming can deplete soil carbon and render soil incapable of supporting life; however, the study also showed that conservation farming can protect carbon in soils, and repair damage over time. The farming practise of cover crops has been recognized as climate-smart agriculture by the White House. In Europe the estimation of the current 0–30 cm SOC stock of agricultural soils was 17.63 Gt. In a subsequent study, authors estimated the best management practices to mitigate soil organic carbon: conversion of arable land to grassland (and vice versa), straw incorporation, reduced tillage, straw incorporation combined with reduced tillage, ley cropping system and cover crops.

An agriculture that mitigates climate change is called Regenerative agriculture. It includes several methods, the main of which are: conservation tillage, diversity, rotation and cover crops, minimizing physical disturbance, minimizing the usage of chemicals. It has other benefits like improving the state of the soil and consequently yields. Some of the big agricultural companies like General Mills and a lot of farms support it.

Societal Controls

Another method being examined is to make carbon a new currency by introducing tradeable "personal carbon credits". The idea being it will encourage and motivate individuals to reduce their 'carbon footprint' by the way they live. Each citizen will receive a free annual quota of carbon that they can use to travel, buy food, and go about their business. It has been suggested that by using this concept it could actually solve two problems; pollution and poverty, old age pensioners will actually be better off because they fly less often, so they can cash in their quota at the end of the year to pay heating bills and so forth.

Population

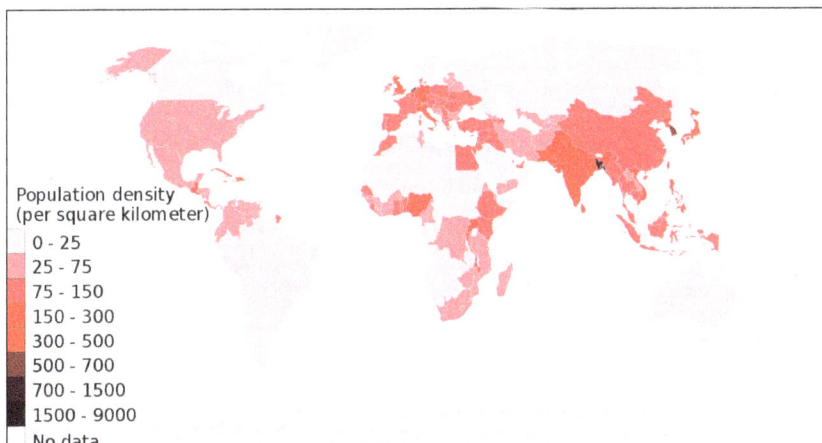

Population density by country.

Various organizations promote population control as a means for mitigating global warming. Proposed measures include improving access to family planning and reproductive health care and information, reducing natalistic politics, public education about the consequences of continued population growth, and improving access of women to education and economic opportunities.

Having one less child would have a much more substantial effect on greenhouse gas emissions compared with for example living car free or eating a plant-based diet. Population control efforts are impeded by there being somewhat of a taboo in some countries against considering any such efforts. Also, various religions discourage or prohibit some or all forms of birth control. Population size has a different per capita effect on global warming in different countries, since the per capita production of anthropogenic greenhouse gases varies greatly by country.

Governmental and Intergovernmental Action

Many countries, both developing and developed, are aiming to use cleaner technologies. Use of these technologies aids mitigation and could result in substantial reductions in CO_2 emissions. Policies include targets for emissions reductions, increased use of renewable energy, and increased energy efficiency. It is often argued that the results of climate change are more damaging in poor nations, where infrastructures are weak and few social services exist. The Commitment to Development Index is one attempt to analyze rich country policies taken to reduce their disproportionate use of the global commons. Countries do well if their greenhouse gas emissions are falling, if their gas taxes are high, if they do not subsidize the fishing industry, if they have a low fossil fuel rate per capita, and if they control imports of illegally cut tropical timber.

Kyoto Protocol

The main current international agreement on combating climate change is the Kyoto Protocol (which is now succeeded by the Paris agreement). On the 11th of December 1997 the Kyoto Protocol was implemented by the 3rd Conference of Parties, which was coming together in Kyoto, which came into force on 16 February 2005. The Kyoto Protocol is an amendment to the United Nations Framework Convention on Climate Change (UNFCCC). Countries that have ratified this protocol have committed to reduce their emissions of carbon dioxide and five other greenhouse gases, or engage in emissions trading if they maintain or increase emissions of these gases. For Kyoto reporting, governments are obliged to be told on the present state of the respective countries' forests and the related ongoing processes.

Temperature Targets

Actions to mitigate climate change are sometimes based on the goal of achieving a particular temperature target. One of the targets that has been suggested is to limit the future increase in global mean temperature (global warming) to below 2 °C, relative to the pre-industrial level. The 2 °C target was adopted in 2010 by Parties to the United Nations Framework Convention on Climate Change. Most countries of the world are Parties to the UNFCCC. The target had been adopted in 1996 by the European Union Council.

Feasibility of 2 °C

Temperatures have increased by 0.8 °C compared to the pre-industrial level, and another

0.5–0.7 °C is already committed. The 2 °C rise is typically associated in climate models with a carbon dioxide equivalent concentration of 400–500 ppm by volume; the current (January 2015) level of carbon dioxide alone is 400 ppm by volume, and rising at 1–3 ppm annually. Hence, to avoid a very likely breach of the 2 °C target, CO_2 levels would have to be stabilised very soon; this is generally regarded as unlikely, based on current programs in place to date. The importance of change is illustrated by the fact that world economic energy efficiency is improving at only half the rate of world economic growth.

Views in the Literature

There is disagreement among experts over whether or not the 2 °C target can be met. For example, according to Anderson and Bows, "there is little to no chance" of meeting the target. On the other hand, according to Alcamo *et al*:

- Policies adopted by parties to the UNFCCC are too weak to meet a 2 or 1.5 °C target. However, these targets might still be achievable if more stringent mitigation policies are adopted immediately.

- Cost-effective 2 °C scenarios project annual global greenhouse gas emissions to peak before the year 2020, with deep cuts in emissions thereafter, leading to a reduction in 2050 of 41% compared to 1990 levels.

Discussion on other Targets

Scientific analysis can provide information on the impacts of climate change and associated policies, such as reducing GHG emissions. However, deciding what policies are best requires value judgements. For example, limiting global warming to 1 °C relative to pre-industrial levels may help to reduce climate change damages more than a 2 °C limit. However, a 1 °C limit may be more costly to achieve than a 2 °C limit.

According to some analysts, the 2 °C "guardrail" is inadequate for the needed degree and timeliness of mitigation. On the other hand, some economic studies suggest more modest mitigation policies. For example, the emissions reductions proposed by Nordhaus might lead to global warming (in the year 2100) of around 3 °C, relative to pre-industrial levels.

Official Long-term Target of 1.5 °C

In 2015, two official UNFCCC scientific expert bodies came to the conclusion that, "in some regions and vulnerable ecosystems, high risks are projected even for warming above 1.5°C". This expert position was, together with the strong diplomatic voice of the poorest countries and the island nations in the Pacific, the driving force leading to the decision of the Paris Conference 2015, to lay down this 1.5 °C long-term target on top of the existing 2 °C goal.

Encouraging use Changes

Emissions Tax

An emissions tax on greenhouse gas emissions requires individual emitters to pay a fee, charge or tax for every tonne of greenhouse gas released into the atmosphere. Most environmentally related taxes with implications for greenhouse gas emissions in OECD countries are levied on energy products and motor vehicles, rather than on CO_2 emissions directly.

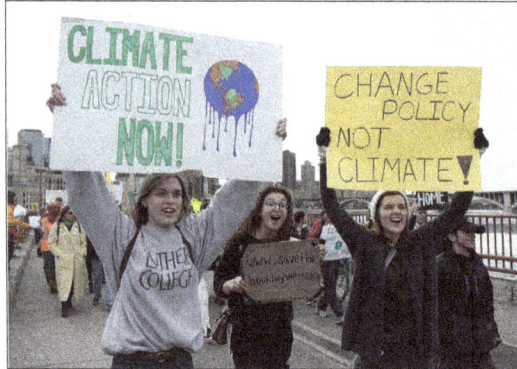

Citizens for climate action at the People's Climate March.

Emission taxes can be both cost-effective and environmentally effective. Difficulties with emission taxes include their potential unpopularity, and the fact that they cannot guarantee a particular level of emissions reduction. Emissions or energy taxes also often fall disproportionately on lower income classes. In developing countries, institutions may be insufficiently developed for the collection of emissions fees from a wide variety of sources.

Subsidies

According to Mark Z. Jacobson, a program of subsidization balanced against expected flood costs could pay for conversion to 100% renewable power by 2030. Jacobson, and his colleague Mark Delucchi, suggest that the cost to generate and transmit power in 2020 will be less than 4 cents per kilowatt hour (in 2007 dollars) for wind, about 4 cents for wave and hydroelectric, from 4 to 7 cents for geothermal, and 8 cents per kWh for solar, fossil, and nuclear power.

Investment

Another indirect method of encouraging uses of renewable energy, and pursue sustainability and environmental protection, is that of prompting investment in this area through legal means, something that is already being done at national level as well as in the field of international investment.

Implementation

Implementation puts into effect climate change mitigation strategies and targets. These can be targets set by international bodies or voluntary action by individuals or institutions. This is the most important, expensive and least appealing aspect of environmental governance.

Funding

Implementation requires funding sources but is often beset by disputes over who should provide funds and under what conditions. A lack of funding can be a barrier to successful strategies as there are no formal arrangements to finance climate change development and implementation. Funding is often provided by nations, groups of nations and increasingly NGO and private sources.

These funds are often channelled through the Global Environmental Facility (GEF). This is an environmental funding mechanism in the World Bank which is designed to deal with global environmental issues. The GEF was originally designed to tackle four main areas: biological diversity, climate change, international waters and ozone layer depletion, to which land degradation and persistent organic pollutant were added. The GEF funds projects that are agreed to achieve global environmental benefits that are endorsed by governments and screened by one of the GEF's implementing agencies.

Problems

There are numerous issues which result in a current perceived lack of implementation. It has been suggested that the main barriers to implementation are: Uncertainty, Fragmentation, Institutional void, Short time horizon of policies and politicians and Missing motives and willingness to start adapting. The relationships between many climatic processes can cause large levels of uncertainty as they are not fully understood and can be a barrier to implementation. When information on climate change is held between the large numbers of actors involved it can be highly dispersed, context specific or difficult to access causing fragmentation to be a barrier. Institutional void is the lack of commonly accepted rules and norms for policy processes to take place, calling into question the legitimacy and efficacy of policy processes. The Short time horizon of policies and politicians often means that climate change policies are not implemented in favour of socially favoured societal issues. Statements are often posed to keep the illusion of political action to prevent or postpone decisions being made. Missing motives and willingness to start adapting is a large barrier as it prevents any implementation. The issues that arise with a system which involves international government cooperation, such as cap and trade, could potentially be improved with a polycentric approach where the rules are enforced by many small sections of authority as opposed to one overall enforcement agency. Concerns about metal requirement and availability for essential decarbonization technoloqies such as photovoltaics, nuclear power, and (plug-in hybrid) electric vehicles have also been expressed as obstacles.

Occurrence

Despite a perceived lack of occurrence, evidence of implementation is emerging internationally. Some examples of this are the initiation of NAPA's and of joint implementation. Many developing nations have made National Adaptation Programs of Action (NAPAs) which are frameworks to prioritize adaption needs. The implementation of many of these is supported by GEF agencies. Many developed countries are implementing 'first generation' institutional adaption plans particularly at the state and local government scale. There has also been a push towards joint implementation between countries by the UNFCCC as this has been suggested as a cost-effective way for objectives to be achieved.

How are Developing Countries approaching Mitigation in Key Sectors?

Most developing countries in Asia and the Pacific find it rather unfair to be asked to contribute to reduction of GHG emissions, especially when their per capita emissions are so much lower than developed countries. Now we live in a globalising world, in which developed and developing countries are intrinsically linked. A significant portion of the production from developing countries is consumed in developed countries, and large volumes of e-waste, for example, are transferred to developing countries for extraction of valuable embedded materials. Indeed, globalisation of industry makes it more difficult to know where ultimate responsibilities for GHG emissions lie. There is always willingness, however, for developing countries as part of a globalised economy to contribute to global mitigation efforts under specific conditions that meet their national economic and social welfare interests.

For example, countries that are interested in maintaining or expanding national forest cover need to find a good economic argument to keep valuable resources "locked up" or to prevent landless farmers and illegal loggers from abrogating state ownership and control of the forest resources. If wealthy developed countries are prepared to pay for carbon sequestration in the forest domain of developing countries, then this can be a win-win situation. The international community is now attempting to extend this win-win logic to "reduced emissions from deforestation and degradation in developing countries" (REDD). As it is not in the interests of the global community to have countries continuously releasing the second largest source of GHG emissions through deforestation, then perhaps developed countries could also pay for the climate benefits of avoided deforestation. This approach would have the added benefits of preserving biodiversity in tropical forests and maintaining critical ecosystem functions, such as protection of watersheds, although there are some technical issues to resolve first. How would such arrangements impinge on forest dependent communities, often with inadequate tenure rights over their traditional forest areas? Would payment for carbon sequestration in the tropical forests of Asia-Pacific assist or detract from sustainable development? Would the inclusion of REDD in global carbon trading schemes have adverse impacts on the price of carbon or should there be a separate market?

A remarkably similar set of policy calculations are, dealing with the controversial issue of biofuels. Here the interests of developing countries lie in creating a new export product and enabling a degree of national energy security. From an economic perspective, countries with abundant land, water and sunlight, plus cheap labour should have a comparative advantage in producing biofuel crops for a rapidly growing global market. Mandatory requirements to achieve certain levels of biofuel use as part of developed country responses to climate change have helped to create a sizable market opening for biodiesel and bioethanol production in developing countries.

In the rush to develop biofuel crops (like oil palm and sugar cane), the impacts of biofuel production on sustainable rural development and food prices are beginning to become clearer. If agricultural land is devoted to biofuel crops, then the opportunity cost of land becomes tied to global energy policy and pricing and is divorced from its critical role in food production and food security. It is also possible that excessive development of biofuel crops may lead to renewed pressure for deforestation, paradoxically linking these two responses to climate change mitigation in a conflicting manner. Oil palm plantations on converted tropical peat land may actually increase GHG emissions rather than contribute to climate change mitigation.

References

- Eccleston, Charles H. (2010). Global Environmental Policy: Concepts, Principles, and Practice. Chapter 7. Isbn 978-1439847664

- Pimentel, berger; et al. (october 2004). "water resources: agricultural and environmental issues". Bioscience. 54 (10): 909. Doi:10.1641/0006-3568(2004)054[0909:wraaei]2.0.co;2

- What-is-environmental-sustainability-and-sustainable-development: conserve-energy-future.com, Retrieved 25 February, 2019

- Sutter, john d.; berlinger, joshua (12 december 2015). "final draft of climate deal formally accepted in paris". Cnn. Cable news network, turner broadcasting system, inc. Retrieved 12 December 201 5

- Newman, varner, lunquist (2018). Defending biodiversity. Cambridge university press. Isbn 9781139024105

Sustainable Development: Theories and Perspectives

A few perspectives and theories related to sustainable development are the tragedy of the commons, eco-development and Gandhian model. These topics explained in this chapter will help in gaining a better perspective about these theories and perspectives of sustainable development.

ECO-DEVELOPMENT

The eco-development prefigures and competes with the expression of the sustainable development. The notion of eco-development has been proposed by Strong in 1972 to reactivate the dialog between the North and the South, after the conference of Stockholm. Sachs, who has occupied a few important charges in the international organs for thirty years, has turned into the principal theoretical of this notion. In The Discovery of the third world, Sachs is conscious of the European-centrism who hides behind the theories elaborated by the economists of the development. Nevertheless, he rejects any abandon of this aim, speaking even about the normative dimension of the eco-development. He prefers qualifying the eco-development of "philosophy of the development" and tries to go to the actors who elaborate concrete projects and intervene in the field.

To Reduce Inequalities

According to Sachs, the economic, including if it is strong and is accompanied by a modernization of the production structures, does not drive to the development. It ends generally in an increase of social inequalities, which are responsible for an important part of the environmental deterioration. A waste takes place when the wealth of some ones drives to the consumption of superfluous products and when the misery of others provokes an overexploitation of the scanty available resources. The "bad-development" is, therefore, a general problem. What supposes starting a global program of reforms in the way of taking the economic decisions, with recommendations differentiated according to the countries, since the responsibilities of the North in the modification of the international relations are specially important. It insists on the necessity of institutional and political changes to allow these evolutions that must take place in a period of three or four generations. Inspired by the conceptions of Kalecki, Sachs does not question the growth but the unequal aims that it provokes. Therefore, his orientation put the economic surplus and the available time, besides work, to the service of social progress and rational management of the natural environments. If the growth is more equal, the population will agree with more pleasure to do restrictions in the satisfaction of their material needs as well as in their demography.

To environmental level, Sachs rejects the hypothesis of an unlimited material and energetic endowment. Nevertheless, the thermodynamic limits assigned to the economic activity by Georgescu-Roegan are valid in a temporary distant horizon. The arrival to the stationary condition, which it will have to resign some day, it is not for immediately. A disagreement exists between the supporters of the eco-development and the defenders of the ecological economy who centre more their attention on the assigned priorities than on the basic problem. Sachs observes that the extensive growth is condemned and his interpretation of the development leaves place to the possibility of a material growth been founded on a more rational use of the energy and matter, which demonstrates that the limits are not given once forever. It allows to Sachs to define the eco-development as a development of the populations by themselves, using the best natural resources, adapting to an environment that it transforms without destroying it.

Autonomy of Decisions and the Types of Development

If the sustainable development appears as a general question in all the countries, it must be declined differently according to the places and the circumstances. The cultural diversity is recognized immediately: all the social groups have the right to continue their aims in the frame of their specific culture and of their relation to the nature. Being opposed to the mimetic strategies of the development, the doctrine of eco-development belongs to the theories of endogenous development. The autonomy of decisions must prevail over the individual and collective levels, insisting the fact that autonomy does not mean autarchy. It is important that every community defines for itself a social global project that appears as his own way of development.

Therefore, it is a question of answering to the following questions: what goods and services must be produced? By who? How? The elements of response are offered by the analysis of the production structure, the revenues and the consumption, as well as by the study of the arrangement of time and space, conducts and values. Definitively, it is a question of a way to be in the world that differs from a site to another and that it is not present immediately to the eyes of the foreign observers. What implies the need to apprehend it correctly, of calling on to the whole range of social sciences and, specially, to history and anthropology. It is a question of reactivating the studies on development in an interdisciplinary perspective, considering the relations between economy and sociology or the links between economy and ecology.

Attention to the Local Development

The circumstantial character of the question of development drives to the ecodevelopment applied locally and regionally. It tries to reveal the specific resources of one "eco-region" with a view to the satisfaction of the fundamental needs of the population. The interest to the eco-development, which is present from the beginning in the rural societies, is reinforced in the urban zones from the 1980's. The regional and local development is conceived as one of the possible routes to overcome the economic and environmental crisis. This reflection about the endogenous development in other territorial scales tries to articulate the fight against the unemployment, the protection of the environment and the summit of the forms of social economy. The eco-development must be translated by a plurality of paths and a model of diversity in a mixed economy.

But, the prudence is necessary and the local development cannot be the panacea. The State appears often as a counterweight to the pressure of local interests and certain environmental questions must be approached in other organization levels. Sachs has remembered constantly the role that the UNO had to recover in the restoration of a new economic international order. With the past of the time, the expectations of a collective management of the common heritage of humanity for a world authority have not weakened, as well as the idea of establishing a world forum to speak about the strategies of development. In this difficult dialectics between the local and global levels, an interrogation takes place on the pertinent spaces of development and on the possibilities to articulate them in an institutional way carried out by the supporters of the eco-development.

Choice of the Appropriate Technologies

Though the problems of development are, first of all, of an institutional and political order, the question of the technical decisions is central in the thought of eco-development. The technology could not be analyzed only with the concepts of capital and work, and his level of abstraction. In the eco-development optics, it is considered like a multidimensional phenomenon that require to take in consideration the type of energy and of resources used to assure his functioning, his results, his complexity, the qualification of the workforce that it needs and his environmental impact. Therefore, it insists on the notion of technologies adapted to the cultural, institutional and ecological context, underlining that are not only intermediate technologies. Breaking with the strategy of mimetic transfer of the technologies of the North towards the South, the appropriate technologies must come from specific investigations that answer to the needs of the countries of the third world and of a selection of the available technologies worldwide. This selectivity must finish with the coexistence of several types of technologies in the bosom of every national economy. The ecological prudence is one of the elements witch be considered in the choice of production technologies.

Participative Planning

The eco-development cannot be reduced to technical choices. It wants to be an essential instrument for the futurology and the exploration of the development options. The State action is traditionally the economic translation of development, knowing that the planning is one of his privileged instruments. The planner of the eco-development must be conversed about the will to optimize his decisions. The optimization is unattainable bearing the multiplicity of the elements in mind to bearing the lack of information relative to a certain number of dimensions of the problems. Therefore, it suits to give him again a political sense to the comprehension of the decision-making processes, the strategies of the actors and the relations of power with which the planner must compose.

The planning is conceived as a place of debate, negotiation and commitment, so that the planner is an entertainer and a negotiator. It appears more as a figure provided with certain qualities than as a system and a few procedures of planning. It exists a multiplicity of dimensions and variables that planner must apprehend and on which it must be able to affect: the distribution of incomes, the structure of consumption, the used technologies, the modalities of utilization of natural resources, the occupation of soil or the exterior trade. To orientate his intervention, this planner must have a series of ecological and social indicators that realize the multidimensionality of the reality. Beyond this withdrawal of information, all the disciplines are summoned to establish the social rationality. It is necessary for a participative planning to allow a just balance between market, State and civil society. This one only receives sense if it accompanies of an effort of education in favour of the populations: if it hopes from them that they could decide about the best conditions and that they could have multiple dimensions, it is indispensable to realize important efforts of formation. Is recognized the need to built a more wide democracy extended to the technical areas.

Environmental Distribution

A tradition in economy of development tries to demonstrate that the development and the absence of development are both faces of the same dynamics of the capitalism, that prosper establishing an interdependence relation between a center and a periphery. The structuralist and marxist theories propose a few theories of the underdevelopment. The economies of the South are open, linked to the international markets, to the evolution of world prices, to the decisions of multinationals, which does that the economic surplus produced is caught by the North economies. These theories do not take in consideration the environmental problematic. This is precisely what Joan Martinez-Alier and his group try to do with the concept of "ecologically unequal exchange" that wide the perspective opened previously.

The exploitation of the South by the North supposes the inequality of wages granted to the workers of the diverse economic spaces. It is allowed that a few equal quantities of work should not be paid to the same price by the actors who take part in the international exchanges. The ecologically unequal exchange describes the fact that certain products are exported by poor countries with prices that do not cover the wage and environmental costs induced by their productions. We can mention the activities of extraction of oil, mineral, transformation of forests in pastures or production of coffee. If Cabeza-Gutés and Martinez-Alier do not reject the international trade as such and prove to be partial to a few more equitable commercial and more respectful exchanges of the environment, some members of the ecological economy call to breaking with the international division of work and to looking, in a pragmatic way, for a major autonomy and even a self-sufficiency of local administration.

Redistributive Challenges of Environmental Questions

Martinez-Alier insists on the problem of ecological distribution. In this way, it places the question of poverty in the center of the challenge of the sustainability. Based on the examples of social movements in the third world countries, he wants to demonstrate that, on one hand, the poverty cannot be considered only a threat for the environment and, on the other hand, that the protection of the environment is not only a luxury reserved to the rich ones. In other terms, he insists on the "ecology of the poor" that fights for a better recognition of his rights, enclosed in the environmental area. This perspective must be taken in consideration since numerous environmental policies provokes conflicts in the North-South relations, it is across the restoration of a permission of negotiable CO_2 emission in case of the prevention forehead the climate change, and it is a route an international trade of genes in the frame of the fight against the erosion of the biodiversity.

Conclusions of the Policies of Regulation

It is still too early to value the effects of the international policies on climate change and biodiversity. These depend on the rules approved about regulation, on initial concession of rights and quotas, and on conditions of his transferability. Guarding in mind that the systems of regulation provided with different rules produce economic and social varied effects, it is interesting to consider the area of fishing that, for more than twenty years, has restored systems of individual transferable quotas to try to regulate the extractions in fishing area.

Towards the Decrease

Toughening even more the debate concerning the notion of sustainable development, some economists propose to support the opposite position with regard to the aim of growth and to restore decrease. These offers have woken a great interest up and have generated alive controversy, even between the supporters of the decrease. Some of them have called on to reject the idea of development, accused of being the mask behind which advances the westernization of the world and the "mercantilisation" of social relations. This position supported by Latouche whose work rejects the development and reconstructs this notion that has a normative content. The sustainable development appears him as a "pretext concept" that development allows to make last. Latouche and the defenders of the postdevelopment propose to replace this aim with that of "lasting decrease".

But, other authors like Harribey criticize the development carried out by the liberal policies and use the notion of sustainable development to defend an alternative model of development. Before restoring a deceleration of growth, the relations of capitalist production must be changed and the inequalities of wealth must be fought, bearing in mind that a period of recovery must be conditioned in order that the populations who need it could see increase his standard of living. In both cases, it is a question of reinventing the imaginary of social change.

The Decrease

The term of decrease is associated with the Georgescu-Roegan's work. His principal merit has been to think about the thermodynamic of the western development. It puts the emphasis in the technical fundamental innovations that have allowed to humanity to use new sources of energy. From this point of view, the human history had only known a few decisive moments: the domestication of fire, the utilization of fossil energies and the succession of coal and oil. The problem of these technologies is that they end up by exhausting the fuel through that they make them live, which he leads to a tragic conception of history of the humanity who is marked by the fights that the individuals face to the States to possess the energetic and material resources. Studying the first machines to steam, in the beginning of the 19th century, the revolution of the productive capacity that they induce modifies the relation of human being to nature.

A thought of the limits is a question itself to object to this immoderation. Georgescu-Roegan is one of the only economists in having recognized the relevancy of the first report of the Club of Rome. He praises the decrease, still being conscious of the basic need to improve the material conditions of the poor populations. It has not stopped remembering that, whenever a car takes place, there are in use quantities of low entropy that might be used to make cars and useful spades for the peasants of the third world. Waiting of hypothetical technologies able of taking the relief of that fossil energies use, it reveals the measures destined to reduce the waste and to minimize the future repentances, allowing that the energetic and material endowments should be possible with time. For it, it calls on to resorting to technical innovations as well as to a straddle of the resources for quantitative instruments that allow starts a strategy of general conservation planned worldwide. Nevertheless, Georgescu-Roegen insists on the need to act on the demand of the products in order to the offer.

Sociableness

For certain aspects, these proposals are alike the critique of the growth elaborated by Illich. Illich's central thesis is that the "religion of growth" legitimizes a technical project that aspires that the industrial manufacture of the existence replaces the invention of the life with individuals. There exist two manners to produce values of use: an autonomous way, for which the individuals answer for themselves to their needs and a heteronomous way that produces goods put to the disposition of persons by the intermediary of a market or of a non-merchant institution. But, due to his efficiency, the heteronomous way has certain trend to be imposed to the autonomous way up to turning into a "racial monopoly", that is to say into a situation where the industrial production destroys any possibility of resorting the other means to satisfy their needs. From this threshold, a counter productivity is observed, to the effect that the institutions end up by producing the opposite of what they should produce. Nevertheless, deprived of his autonomy, coffee with a dash of milk of the others and of the world, the individual does not have other possibilities that of going to the industry, which reduces even more his autonomy and reinforces the obligation to consume services produced industrially. The search of well-being drives to a loss of control increasingly big of his existence on the part of the individuals. Before this evolution, Illich calls on the individuals to take again the control of their lives and built a sociable society, where the persons control the instruments of their environment.

A Norm of what is Sufficient

This search of autonomy of the individuals drives to consider equally in a critical way the historical and psycho-sociological bows that join the productivism, the consumerism and the organization of work. Gorz remembers that the first manufacturers have had difficulties to achieve on the part of the workers a constant, regular and complete work, in spite of promising them higher salaries. Before, these workers were working the time that it was necessary to attend to their needs. This limitation of the needs was allowing a selflimitation of the effort of each one and the work of all. Based on the possibilities offered by the technology, it has dispossessed the workers of the instruments of production, of the product of their work and of the work itself, in order that the production could become emancipated from the sufficiency.

The invention of the factory has allowed the modification of the relation to the nature and the empowerment of the capitalist on the productive process. It has diminished the salary of workers in order that they work more than the necessary and, little by little, a disjunction has been restored between the labor time and the private time. The loss of sense has established itself, provided that the work is lived by the majority of these individuals as the way of gaining a salary. In parallel, one has been present at the creation of an increasing number of needs to satisfy, since the individuals buy certain products for lack of time to be able to realize these tasks themselves. The merchant consumption has increased equally with the game of a phenomenon of existential compensation.

The exit of this dynamics forces to accept a few resignations. The current challenge according to Gorz is to restore politically a norm of what is sufficient relatively to the contemporary living conditions. This reduction of the merchant consumption, this decrease of the economy happens for a different distribution from the improvements of productivity and a reduction of the time work, conceived as a long-term politics, always and when a sufficient revenue is guaranteed independently of the duration of work and that produces a redistribution of work to himself so that the whole world could work less and better. This liberated time must allow the autonomy of the individuals, the auto-production, the constitution of networks, of solidarities, of cooperation and of investments in the political area.

TRAGEDY OF THE COMMONS

The tragedy of the commons is a situation in a shared-resource system where individual users, acting independently according to their own self-interest, behave contrary to the common good of all users, by depleting or spoiling that resource through their collective action. The theory originated in an essay written in 1833 by the British economist William Forster Lloyd, who used a hypothetical example of the effects of unregulated grazing on common land (also known as a "common") in Great Britain and Ireland. The concept became widely known as the "tragedy of the commons" over a century later. In this modern economic context, "commons" is taken to mean any shared and unregulated resource such as atmosphere, oceans, rivers, fish stocks, roads and highways, or even an office refrigerator.

Cows on Selsley Common. The tragedy of the commons is one way of accounting for overexploitation.

The term is used in environmental science. The "tragedy of the commons" is often cited in connection with sustainable development, meshing economic growth and environmental protection, as well as in the debate over global warming. It has also been used in analyzing behavior in the fields of economics, evolutionary psychology, anthropology, game theory, politics, taxation and sociology.

Although common resource systems have been known to collapse due to overuse (such as in over-fishing), many examples have existed and still do exist where members of a community with access to a common resource co-operate or regulate to exploit those resources prudently without collapse. Elinor Ostrom was awarded the 2009 Nobel Prize in economics for demonstrating exactly this concept in her book Governing the Commons, which included examples of how local communities were able to do this without top-down regulations.

It has been argued that the very term "tragedy of the commons" is a misnomer since "the commons" referred to land resources with rights jointly owned by members of a community, and no individual outside the community had any access to the resource. However, the term is now used in social science and economics when describing a problem where all individuals have equal and open access to a resource. Hence, "tragedy of open access regimes" or simply "the open access problem" are more apt terms.

Expositions

Lloyd's Pamphlet

In 1833, the English economist William Forster Lloyd published a pamphlet which included a hypothetical example of over-use of a common resource. This was the situation of cattle herders sharing a common parcel of land on which they are each entitled to let their cows graze, as was the custom in English villages. He postulated that if a herder put more than his allotted number of cattle on the common, overgrazing could result. For each additional animal, a herder could receive additional benefits, but the whole group shared damage to the commons. If all herders made this individually rational economic decision, the common could be depleted or even destroyed, to the detriment of all.

Garrett Hardin's

In 1968, ecologist Garrett Hardin explored this social dilemma in his article "The Tragedy of the Commons", published in the journal *Science*. The essay derived its title from the pamphlet by Lloyd, which he cites, on the over-grazing of common land.

Hardin discussed problems that cannot be solved by technical means, as distinct from those with solutions that require "a change only in the techniques of the natural sciences, demanding little or nothing in the way of change in human values or ideas of morality". Hardin focused on human population growth, the use of the Earth's natural resources, and the welfare state. Hardin argued that if individuals relied on themselves alone, and not on the relationship of society and man, then the number of children had by each family would not be of public concern. Parents breeding excessively would leave fewer descendants because they would be unable to provide for each child adequately. Such negative feedback is found in the animal kingdom. Hardin said that if the children of improvident parents starved to death, if overbreeding was its own punishment, then there would be no public interest in controlling the breeding of families. Hardin blamed the welfare state for allowing the tragedy of the commons; where the state provides for children and supports overbreeding as a fundamental human right, Malthusian catastrophe is inevitable. Consequently, in his article, Hardin lamented the following proposal from the United Nations:

> The Universal Declaration of Human Rights describes the family as the natural and fundamental unit of society. It follows that any choice and decision with regard to the size of the family must irrevocably rest with the family itself, and cannot be made by anyone else.

In addition, Hardin also pointed out the problem of individuals acting in rational self-interest by claiming that if all members in a group used common resources for their own gain and with no regard for others, all resources would still eventually be depleted. Overall, Hardin argued against relying on conscience as a means of policing commons, suggesting that this favors selfish individuals – often known as free riders – over those who are more altruistic.

In the context of avoiding over-exploitation of common resources, Hardin concluded by restating Hegel's maxim (which was quoted by Engels), "freedom is the recognition of necessity". He suggested that "freedom" completes the tragedy of the commons. By recognizing resources as commons in the first place, and by recognizing that, as such, they require management, Hardin believed that humans "can preserve and nurture other and more precious freedoms".

Commons as a Modern Resource Concept

Hardin's article was the start of the modern use of "Commons" as a term connoting a shared resource. As Frank van Laerhoven & Elinor Ostrom have stated: "Prior to the publication of Hardin's article on the tragedy of the commons, titles containing the words 'the commons', 'common pool resources,' or 'common property' were very rare in the academic literature." They go on to say: "In 2002, Barrett and Mabry conducted a major survey of biologists to determine which publications in the twentieth century had become classic books or benchmark publications in biology. They report that Hardin's 1968 article was the one having the greatest career impact on biologists and is the most frequently cited".

Application

Metaphoric Meaning

Like Lloyd and Thomas Malthus before him, Hardin was primarily interested in the problem of human population growth. But in his essay, he also focused on the use of larger (though finite) resources such as the Earth's atmosphere and oceans, as well as pointing out the "negative commons" of pollution (i.e., instead of dealing with the deliberate privatization of a positive resource, a "negative commons" deals with the deliberate commonization of a negative cost, pollution).

As a metaphor, the tragedy of the commons should not be taken too literally. The "tragedy" is not in the word's conventional or theatric sense, nor a condemnation of the processes that lead to it. Similarly, Hardin's use of "commons" has frequently been misunderstood, leading him to later remark that he should have titled his work "The Tragedy of the Unregulated Commons".

The metaphor illustrates the argument that free access and unrestricted demand for a finite resource ultimately reduces the resource through over-exploitation, temporarily or permanently. This occurs because the benefits of exploitation accrue to individuals or groups, each of whom is motivated to maximize use of the resource to the point in which they become reliant on it, while the costs of the exploitation are borne by all those to whom the resource is available (which may be a wider class of individuals than those who are exploiting it). This, in turn, causes demand for the resource to increase, which causes the problem to snowball until the resource collapses (even if it retains a capacity to recover). The rate at which depletion of the resource is realized depends primarily on three factors: the number of users wanting to consume the common in question, the consumptiveness of their uses, and the relative robustness of the common.

The same concept is sometimes called the "tragedy of the fishers", because fishing too many fish before or during breeding could cause stocks to plummet.

Modern Commons

The *tragedy of the commons* can be considered in relation to environmental issues such as sustainability. The commons dilemma stands as a model for a great variety of resource problems in society today, such as water, forests, fish, and non-renewable energy sources such as oil and coal.

Situations exemplifying the "tragedy of the commons" include the overfishing and destruction of the Grand Banks, the destruction of salmon runs on rivers that have been dammed – most prominently in modern times on the Columbia River in the Northwest United States, and historically in North Atlantic rivers – the devastation of the sturgeon fishery – in modern Russia, but historically in the United States as well – and, in terms of water supply, the limited water available in arid regions (e.g., the area of the Aral Sea) and the Los Angeles water system supply, especially at Mono Lake and Owens Lake.

In economics, an externality is a cost or benefit that affects a party who did not choose to incur that cost or benefit. Negative externalities are a well-known feature of the "tragedy of the commons". For example, driving cars has many negative externalities; these include pollution, carbon emissions, and traffic accidents. Every time 'Person A' gets in a car, it becomes more likely that 'Person Z' – and millions of others – will suffer in each of those areas. Economists often urge the government to adopt policies that "internalize" an externality.

Examples:

More general examples (some alluded to by Hardin) of potential and actual tragedies include:

Clearing rainforest for agriculture in southern Mexico.

- Planet Earth ecology:
  - Uncontrolled human population growth leading to overpopulation.
  - A preference for sons made people abort foetal girls. This results in an imbalanced gender ratio.
  - Atmosphere, through the release of pollution that leads to Ozone depletion, global warming, ocean acidification (by way of increased atmospheric CO_2 being absorbed by the sea), and particulate pollution.
  - Indoor air.
  - Light pollution: with the loss of the night sky for research and cultural significance, affected human, flora and fauna health, nuisance, trespass and the loss of enjoyment or function of private property.
  - Water: Water pollution, water crisis of over-extraction of groundwater and wasting water due to overirrigation.

- ○ Forests: Frontier logging of old growth forest and slash and burn.

- ○ Energy resources and climate: Environmental residue of mining and drilling, Burning of fossil fuels and consequential global warming.

- ○ Animals: Habitat destruction and poaching leading to the Holocene mass extinction

- ○ Human and wildlife conflict.

- ○ Oceans: Overfishing.

- ○ Antibiotics: Antibiotic Resistance Misuse of antibiotics anywhere in the world will eventually result in antibiotic resistance developing at an accelerated rate. The resulting antibiotic resistance has spread (and will likely continue to do so in the future) to other bacteria and other regions, hurting or destroying the Antibiotic Commons that is shared on a worldwide basis.

- Publicly shared resources:

  - ○ Spam email degrades the usefulness of the email system and increases the cost for all users of the Internet while providing a benefit to only a tiny number of individuals.

  - ○ Wi-Fi and its overcrowded 2.4 Ghz channels.

  - ○ Vandalism and littering in public spaces such as parks, recreation areas, and public restrooms.

  - ○ Knowledge commons encompass immaterial and collectively owned goods in the information age, including, for example:

    - ▪ Source code and software documentation in software projects that can get "polluted" with messy code or inaccurate information.

    - ▪ Skills acquisition and training, when all parties involved pass the buck on implementing it.

  - ○ Electric vehicle (EV) charging station blocked by parked vehicles, ICE vehicles whose drivers resent EVs, EVs that overstay time limits, and EVs whose owners have no intention of charging but feel they are entitled to park.

Application to Evolutionary Biology

A parallel was drawn recently between the tragedy of the commons and the competing behaviour of parasites that through acting selfishly eventually diminish or destroy their common host. The idea has also been applied to areas such as the evolution of virulence or sexual conflict, where males may fatally harm females when competing for matings. It is also raised as a question in studies of social insects, where scientists wish to understand why insect workers do not undermine the "common good" by laying eggs of their own and causing a breakdown of the society.

The idea of evolutionary suicide, where adaptation at the level of the individual causes the whole species or population to be driven extinct, can be seen as an extreme form of an evolutionary tragedy of the commons. From an evolutionary point of view, the creation of the tragedy of the commons in pathogenic microbes may provide us with advanced therapeutic methods.

Commons Dilemma

The *commons dilemma* is a specific class of social dilemma in which people's short-term selfish interests are at odds with long-term group interests and the common good. In academia, a range of related terminology has also been used as shorthand for the theory or aspects of it, including *resource* dilemma, take-some dilemma, and common pool resource.

Commons dilemma researchers have studied conditions under which groups and communities are likely to under- or over-harvest common resources in both the laboratory and field. Research programs have concentrated on a number of motivational, strategic, and structural factors that might be conducive to management of commons.

In game theory, which constructs mathematical models for individuals' behavior in strategic situations, the corresponding "game", developed by Hardin, is known as the Commonize Costs – Privatize Profits Game (CC–PP game).

Psychological Factors

Kopelman, Weber, & Messick, in a review of the experimental research on cooperation in commons dilemmas, identify nine classes of independent variables that influence cooperation in commons dilemmas: social motives, gender, payoff structure, uncertainty, power and status, group size, communication, causes, and frames. They organize these classes and distinguish between psychological individual differences (stable personality traits) and situational factors (the environment). Situational factors include both the task (social and decision structure) and the perception of the task.

Empirical findings support the theoretical argument that the cultural group is a critical factor that needs to be studied in the context of situational variables. Rather than behaving in line with economic incentives, people are likely to approach the decision to cooperate with an appropriateness framework. An expanded, four factor model of the Logic of Appropriateness, suggests that the cooperation is better explained by the question: "What does a person like me (identity) do (rules) in a situation like this (recognition) given this culture (group)?"

Strategic Factors

Strategic factors also matter in commons dilemmas. One often-studied strategic factor is the order in which people take harvests from the resource. In simultaneous play, all people harvest at the same time, whereas in sequential play people harvest from the pool according to a predetermined sequence – first, second, third, etc. There is a clear order effect in the latter games: the harvests of those who come first – the leaders – are higher than the harvest of those coming later – the followers. The interpretation of this effect is that the first players feel entitled to take more. With sequential play, individuals adopt a first come-first served rule, whereas with simultaneous play people may adopt an equality rule. Another strategic factor is the ability to build up reputations.

Research found that people take less from the common pool in public situations than in anonymous private situations. Moreover, those who harvest less gain greater prestige and influence within their group.

Structural Factors

Much research has focused on when and why people would like to structurally rearrange the commons to prevent a tragedy. Hardin stated in his analysis of the tragedy of the commons that "Freedom in a commons brings ruin to all." One of the proposed solutions is to appoint a leader to regulate access to the common. Groups are more likely to endorse a leader when a common resource is being depleted and when managing a common resource is perceived as a difficult task. Groups prefer leaders who are elected, democratic, and prototypical of the group, and these leader types are more successful in enforcing cooperation. A general aversion to autocratic leadership exists, although it may be an effective solution, possibly because of the fear of power abuse and corruption.

The provision of rewards and punishments may also be effective in preserving common resources. Selective punishments for overuse can be effective in promoting domestic water and energy conservation – for example, through installing water and electricity meters in houses. Selective rewards work, provided that they are open to everyone. An experimental carpool lane in the Netherlands failed because car commuters did not feel they were able to organize a carpool. The rewards do not have to be tangible. In Canada, utilities considered putting "smiley faces" on electricity bills of customers below the average consumption of that customer's neighborhood.

Articulating solutions to the tragedy of the commons is one of the main problems of political philosophy. In many situations, locals implement (often complex) social schemes that work well. The best governmental solution may be to do nothing. When these fail, there are many possible governmental solutions such as privatization, internalizing the externalities, and regulation.

Non-governmental Solution

Sometimes the best governmental solution may be to do nothing. Robert Axelrod contends that even self-interested individuals will often find ways to cooperate, because collective restraint serves both the collective and individual interests. Anthropologist G. N. Appell criticized those who cited Hardin to "impose their own economic and environmental rationality on other social systems of which they have incomplete understanding and knowledge."

Political scientist Elinor Ostrom, who was awarded 2009's Nobel Memorial Prize in Economic Sciences for her work on the issue, and others revisited Hardin's work in 1999. They found the tragedy of the commons not as prevalent or as difficult to solve as Hardin maintained, since locals have often come up with solutions to the commons problem themselves. For example, it was found that a commons in the Swiss Alps has been run by a collective of farmers there to their mutual and individual benefit since 1517, in spite of the farmers also having access to their own farmland. In general, it is in the interest of the users of a commons to keep them functioning and so complex social schemes are often invented by the users for maintaining them at optimum efficiency.

Similarly, geographer Douglas L. Johnson remarks that many nomadic pastoralist societies of Africa and the Middle East in fact "balanced local stocking ratios against seasonal rangeland conditions in ways that were ecologically sound", reflecting a desire for lower risk rather than higher profit; in spite of this, it was often the case that "the nomad was blamed for problems that were not of his own making and were a product of alien forces." Independently finding precedent in the opinions of previous scholars such as Ibn Khaldun as well as common currency in antagonistic cultural attitudes towards non-sedentary peoples, governments and international organizations have made use of Hardin's work to help justify restrictions on land access and the eventual sedentarization of pastoral nomads despite its weak empirical basis. Examining relations between historically nomadic Bedouin Arabs and the Syrian state in the 20th century, Dawn Chatty notes that "Hardin's argument was curiously accepted as the fundamental explanation for the degradation of the steppe land" in development schemes for the arid interior of the country, downplaying the larger role of agricultural overexploitation in desertification as it melded with prevailing nationalist ideology which viewed nomads as socially backward and economically harmful.

Elinor Ostrom and her colleagues looked at how real-world communities manage communal resources, such as fisheries, land irrigation systems, and farmlands, and they identified a number of factors conducive to successful resource management. One factor is the resource itself; resources with definable boundaries (e.g., land) can be preserved much more easily. A second factor is resource dependence; there must be a perceptible threat of resource depletion, and it must be difficult to find substitutes. The third is the presence of a community; small and stable populations with a thick social network and social norms promoting conservation do better. A final condition is that there be appropriate community-based rules and procedures in place with built-in incentives for responsible use and punishments for overuse. When the commons is taken over by non-locals, those solutions can no longer be used.

Governmental Solutions

Governmental solutions may be necessary when the above conditions are not met (such as a community being too big or too unstable to provide a thick social network). Examples of government regulation include privatization, regulation, and internalizing the externalities.

Privatization

One solution for some resources is to convert common good into private property, giving the new owner an incentive to enforce its sustainability. Libertarians and classical liberals cite the tragedy of the commons as an example of what happens when Lockean property rights to homestead resources are prohibited by a government. They argue that the solution to the tragedy of the commons is to allow individuals to take over the property rights of a resource, that is, to privatize it. In England, this solution was attempted in the Inclosure Acts.

Regulation

In a typical example, governmental regulations can limit the amount of a common good that is available for use by any individual. Permit systems for extractive economic activities including mining, fishing, hunting, livestock raising and timber extraction are examples of this approach. Similarly, limits to pollution are examples of governmental intervention on behalf of the commons.

This idea is used by the United Nations Moon Treaty, Outer Space Treaty and Law of the Sea Treaty as well as the UNESCO World Heritage Convention which involves the international law principle that designates some areas or resources the Common Heritage of Mankind.

In Hardin's essay, he proposed that the solution to the problem of overpopulation must be based on "mutual coercion, mutually agreed upon" and result in "relinquishing the freedom to breed". Hardin discussed this topic further in a 1979 book, Managing the Commons, co-written with John A. Baden. He framed this prescription in terms of needing to restrict the "reproductive right", to safeguard all other rights. Several countries have a variety of population control laws in place.

German historian Joachim Radkau thought Hardin advocates strict management of common goods via increased government involvement or international regulation bodies. An asserted impending "tragedy of the commons" is frequently warned of as a consequence of the adoption of policies which restrict private property and espouse expansion of public property.

Internalizing Externalities

Privatization works when the person who owns the property (or rights of access to that property) pays the full price of its exploitation. As discussed above negative externalities (negative results, such as air or water pollution, that do not proportionately affect the user of the resource) is often a feature driving the tragedy of the commons. *Internalizing the externalities*, in other words ensuring that the users of resource pay for all of the consequences of its use, can provide an alternate solution between privatization and regulation. One example is gasoline taxes which are intended to include both the cost of road maintenance and of air pollution. This solution can provide the flexibility of privatization while minimizing the amount of government oversight and overhead that is needed.

Criticism

Radical environmentalist Derrick Jensen claims the tragedy of the commons is used as propaganda for private ownership. He says it has been used by the political right wing to hasten the final enclosure of the "common resources" of third world and indigenous people worldwide, as a part of the Washington Consensus. He argues that in true situations, those who abuse the commons would have been warned to desist and if they failed would have punitive sanctions against them. He says that rather than being called "The Tragedy of the Commons", it should be called "the Tragedy of the Failure of the Commons".

Marxist geographer David Harvey has a similar criticism, noting that "The dispossession of indigenous populations in North America by 'productive' colonists, for instance, was justified because indigenous populations did not produce value", and asks generally: "Why, for instance, do we not focus in Hardin's metaphor on the individual ownership of the cattle rather than on the pasture as a common?"

Hardin's work was also criticised as historically inaccurate in failing to account for the demographic transition, and for failing to distinguish between common property and open access resources. In a similar vein, Carl Dahlman argues that commons were effectively managed to prevent overgrazing. Likewise, Susan Jane Buck Cox argues that the common land example used to argue this economic concept is on very weak historical ground, and misrepresents what she terms was actually the "triumph of the commons": the successful common usage of land for many centuries. She argues that social changes and agricultural innovation, and not the behaviour of the commoners, led to the demise of the commons.

Some authors, like Yochai Benkler, say that with the rise of the Internet and digitalisation, an economics system based on commons becomes possible again. He wrote in his book *The Wealth of Networks* in 2006 that cheap computing power plus networks enable people to produce valuable products through non-commercial processes of interaction: "as human beings and as social beings, rather than as market actors through the price system". He uses the term *networked information economy* to refer to a "system of production, distribution, and consumption of information goods characterized by decentralized individual action carried out through widely distributed, nonmarket means that do not depend on market strategies." He also coined the term *commons-based peer production* for collaborative efforts based on sharing information.

GANDHIAN MODEL OF SUSTAINABLE DEVELOPMENT

The theme of sustainable development has evolved with the evolution of human civilization. The very beginning of human society and its onward march is woven critically around this concept which has assumed significance for the survival of the modern civilization and planet earth. Whenever human civilization receded from the path of sustainable development the danger to its survival was ensured. With the advent of industrial revolution in Europe began the era of unsustainable development. The unleashing of creative energies of people during that period led to the spectacular progress in the field of science and technology. The tapping of energy from coal and the application of new methods of production gave rise to unprecedented productivity.

While industrial revolution released humans from the fetters of feudalism and bigotry it put new chains around them in the form of materialism and materialistic appetite. The mind which became free from bondage of bigotry and exploitative feudal mode of production became subservient to machinery and greed. Driven by the credo of mass production the modern western civilization chose the path of violence subjugating the territories of the peoples of Asia, Africa and Latin America and appropriating their resources. Conquest and exploitation of the human and material resources beyond the boundaries of Europe became the guiding aspect of that civilization. The policies and values associated with that path led to the indiscriminate consumption of energy and resources of the planet earth and gave birth to an imperial mindset.

By 1980s it was realized that such an approach would degrade the environment beyond repair and cause unimaginable consequences to the very existence of the planet. An institutionalized approach in the form of The World Commission on Environment and Development under the Chairmanship of Harlem Brundtland was set up to find remedies to the problem. It produced a report in 1987 entitled "Our Common Future" which stressed on the ability of mankind to make development sustainable.

It understood sustainable development in terms of "the limits imposed by the present state of technology and social organizations on environmental resources and by the ability of the biosphere to absorb the effects of human activities." A pertinent question exercises our mind. Why was the Commission established in 1983 and not before? A provisional answer was that by the beginning of the 1980s it was painfully realized that the western world was living beyond the limits of the planet earth. "The Living Planet" a report of the World Wild Life Foundation released in 2006 clearly stated that "in 1980s it was realized that the Humanity's footprint first grew larger than global biocapacity" disturbing the subtle balance of the planet earth. The Human Development Report 2007-08 on the theme Fighting Climate Change: Human Solidarity in a Divided World also critically looked at the modern development and wrote "Climate change calls into question the enlightenment principle that human progress will make the future look better than the past."

It is indeed tragic that it took so many centuries to realize that the mode of life of the western people and their path of development was unsustainable and therefore an attempt was made to search for the ways and means to rectify the course. There is in fact a desperate search for methods to decarbonise the environment and reestablish the atmosphere prevailing before industrial revolution.

It is in this context that one is struck by the approach of Mahatma Gandhi who in the first decade of the twentieth century understood the unsustainability of the modern civilization based on multiplication of wants and desires. He launched the first Satyagraha in 1906 which lasted for eight years and ended in 1914. Based on truth and non-violence it stressed on simplicity of life, unity of all religions and of the entire mankind. The Common Future which Europeans understood through Brundtland Commission in 1987 was understood by Mahatma Gandhi in the end of 19th century itself. Through his book "The Hind Swaraj" he outlined the threat to common future of humanity caused by relentless quest for more material goods and services. He described the civilization driven by endless multiplication of wants as Satanic and defined civilization in terms of performance of duties, adherence to morality and exercise of restraint. Any approach which puts limitations on passion and greed and which aims at fulfilling the fundamental needs remain central to the concept and practice of sustainable development.

The Hind Swaraj became a manifesto of sustainable development. Even though it did not refer to nature or environment in any of its passages it exposed the predatory instincts of modern civilization and thereby became an important publication critically scrutinizing the modern civilization which was at its zenith. It prophetically stated that modern urban industrial civilization contained in itself its own seeds of destruction. Hind Swaraj was a remarkable outcome of the first Satyagraha. Therefore his first Satyagraha launched for restoration of the democratic rights of Indians became a Satyagraha against the exploitation of the modern western civilization. In a much broader sense it was a Satyagraha which had the challenging and compassionate vision of saving the planet earth.

Tackling air pollution by adopting suitable remedial measures is one of the essential aspects of sustainability and sustainable development. It is educative to note that Mahatma Gandhi while spearheading the first Satyagraha in South Africa in 1913 observed that in modern civilisation access to clean air involved some cost and expenses. In his illuminating write up "Key to Health" which had a separate chapter on Air he observed that the structure of the body needed three kinds of nourishment: air, water and food and of these air constituted the most essential aspect. Stating that "Nature has provided it to such extent that we can have it at no cost" he noted with anguish. "But modern civilization has put a price even on air. In these times, one has to go off to distant places to take the air, and this costs money". On 1st January 1918, hundred years back, he while addressing a meeting in Ahmedabad defined independence of India in terms of three elements- Air, Water and Grain. What he did in 1918 is being done by law courts to explain right to life in terms of right to clean air and water and adequate food. Yet again in late 1930s he defined democracy in terms of access of all citizens to pure air and water. All such understanding of Gandhi on air more than hundred years back and his contextualisation of clean air by referring to modern civilisation, independence of India and democracy make his ideas so contemporary for twenty first century world grappling to uphold sustainability wholly, substantially and in full measure.

It is of paramount importance to note that sustainable development implying harmonious existence of mankind with nature and ecology presupposed an approach based on equity and justice and coexistence of all cultures and civilizations. An unsustainable path of development centering around domination and conquest of other peoples and their natural resources give rise to an imperial world view which detects "fault lines" along nations and cultures and views the existence of different races in antagonistic terms. Towards the end of the twentieth century the celebrated American Scholar Samuel Huntington came out with the theory of clash of civilizations. His hypothesis in its expanded form assumed that diverse civilizations in the world would clash substituting the wars among nations. The dangerous thesis is a byproduct of the modern civilization which emerged after the industrial revolution and which contemptuously treated the civilizations and cultures of peoples of Asia, Africa and Latin America. What Samuel Huntington wrote towards the end of the twentieth century was advocated by General Smuts in the beginning of the same century. When Indians fought non-violently against the restrictions imposed on them for their entry into Transvaal, he wrote to Mahatma Gandhi "South Africa is a representative western civilization while India is the centre of oriental culture. Thinkers of the present generation hold that these two civilizations cannot go together. If nations representing these rival cultures meet even in small groups, the result will only be an explosion."

The incompatibility of the oriental and western civilizations outlined by General Smuts underlined the incompatibility of approaches of the two civilizations – the former stressing on simplicity and restraint and the later on extravagant consumption pattern. He in fact wrote in so many words that people of the oriental culture with their simple habits and contended life have an outlook which run contrary to the outlook of people belonging to western culture which taught them to have and possess more and if necessary to shed blood for achieving that goal. What General Smuts wrote to Gandhiji was nothing but a gross and unabashed manifestation of an extravagant life style unmindful of its consequences on nature and other sections of humanity.

Mahatma Gandhi through his first Satyagraha in South Africa and subsequently during our freedom movement was engaged in criticizing the colonial modernity which went beyond the carrying capacity of the planet earth and exploited peoples and resources across the planet. Therefore our freedom struggle under his leadership was in a way the first ever struggle in history for sustainable development. There are many statements of Mahatma Gandhi which can be quoted to substantiate this point. One particular statement he made in the context of Europeans is of abiding relevance for the whole mankind. He wrote in 1931:

> "The incessant search for material comfort and their multiplication is such an evil and I make bold to say that the Europeans themselves will have to remodel their outlook, if they are not to perish under the weight of the comforts to which they are becoming slaves."

In fact the Europeans are gradually listening to the ringing words of Mahatma Gandhi. It is evident from the approach of some British citizens who have taken measures to simplify life so as to reduce their dependence on energy and resources. They have established a zero-energy (fossil) development system which enables them to run a housing society in London. At the entrance of the Society there is a line written which reads as follows:

> "If every one on the planet consumed as much as the average person in the U.K. we wood need three planets to support us."

These words recapture thoughts of Gandhiji who eight decades back wrote that if India followed the western model of development she would require more than one planet to achieve the progress they had attained.

The residents of the Housing Society in no way belong to the movements launched to protect climate and environment. They pursue diverse professions and services and are a part of the vibrant middle class. What distinguishes them is their remodeled outlook which eschew excessive consumption and production and follow in practice the methods of simple living.

They have resolved not to eat food which come from distant places. They are convinced that when items are transported from long distances a lot of energy is used for transporting, preserving and packing them. The growing consciousness that dependence on food from far off places would lead to excessive use of energy which in turn would lead to emission of more carbon dioxide and green house gases persuades them to use resources available within a few kilometers.

The Nicolas Stern Committee Report from the U.K. on Economics of Climate Change also underlined the same point when it observed that at the current rate of consumption of resources and energy of the planet, mankind would require more than one planet for survival. The Stern Committee Report therefore stressed on reduction of green house gas emissions by remodeling life style and by transiting from a carbon economy to a non-carbon economy.

What is being done in the ZED Housing Society and what is being recommended by the Nicolas Stern Committee Report was highlighted by Mahatma Gandhi during his first Satyagraha and our freedom struggle. He wrote on numerous occasions that failure on the part of human beings to satisfy their material needs by using resources available with fifteen or twenty kilometers would disturb the economy of nature. His usage of the phrase "economy of nature" in 1911 brings out his sensitivity and deeper understanding of human action vis-à-vis ecology.

In the Hind Swaraj he wrote against the annihilation of distance and time. While doing so he did not refer to the danger of excessive energy consumption. Yet with his remarkable intuitive understanding of the danger of modern technology on society and nature he advised mankind to simplify life. His critique of modern civilization, his condemnation of attempts to annihilate distance and his own life of simple and restrained living constituted refreshing attempts to establish a non-carbon and non-exploitative economy in the world.

In earlier paras we had argued that Mahatma Gandhi through our freedom struggle was going beyond the issue of independence and critically evaluating the colonial modernity which violently appropriated the resources of the planet and caused untold misery to the peoples in Asia, Africa and Latin America. The imperial rule and mindset was a byproduct of unsustainable development which was based on exploitation and injustice. He wanted India to avoid that path for the sake of not only Indians but for the whole mankind. Today people in Europe, as mentioned above, are realizing that their style and pattern of life cannot be sustained by utilizing resources available in our own planet. Their life style is a continuation of the colonial mindset based on their presumption that nature and planet earth has limitless resources and they have the right to use them to the exclusion of the rights of others. The world view which excludes others and exploits their resources for the benefit of chosen few is a dangerous and unsustainable world view. As early as 1894 Mahatma Gandhi had written that the policy of exclusion has become obsolete. In subsequent decades he outlined the pain and misery caused by such a world view and cautioned that if India followed that approach it would spell incalculable danger to the whole mankind. His insights were reflected in a small passage on Indusrialism which he authored in 1928.

The world of twenty-first century is reaping the adverse consequences of the industrialization process which was set in motion by the western nations after industrial revolution. It has become vulnerable to unimaginable destruction due to a development process which brought severe pressure on natural resources and which is almost depleting the finite reservoir of energy derived from hydrocarbon. Mahatma Gandhi's early warning in the form of the above statement of 1928 sounds so contemporary for a world confronting the unprecedented danger of global warming and climate change.

Mahatma Gandhi's outstanding leadership during India's struggle for independence was a leadership for a sustainable world order. He spoke, wrote and put into practice many ideas which brought out his leadership qualities for the cause of sustainable development. We are familiar with his historic Dandi March of 1930. The very reference to that March stirs our mind in grasping his unprecedented method of asserting the right of common people over natural resources of which salt is the most basic and primary one. The British empire thrived in monopolizing resources and depriving their legitimate owners access to them. Denial of access of common people to the basic resources is part of a strategy for unsustainable development. Mahatma Gandhi by breaking the salt law and asserting the rights of ordinary people to make salt was empowering the common people which is central to the issue of sustainable development.

After Dandi March was over, he outlined its larger goal by stating that the aim of the March went beyond the independence of India and encompassed in its scope the much broader objective of freeing the world from the monstrous greed of materialism. It was a powerful statement which in combination with his criticism of the greed based modern civilization made Mahatma Gandhi one of the greatest exponents and practitioners of sustainable development. In fact Joseph Stiglitz in his latest book 'Making Globalisation Work' wrote that in a globalised world the western nations give precedence to material values over environmental values. Mahatma Gandhi was once told by a British correspondent that in a materialistic world non-violence would not be effective. In responding to that observation Gandhiji wrote that when non-violence reigned supreme materialism would take a back seat. Through Dandi March and indeed through his path-breaking non-violent work beginning in South Africa and culminating with his martyrdom he wanted non-violence to reign supreme. Creatively interpreting non-violence and non-violent mass action in its broadest sense he stressed, among other things, on communal harmony, economic equality, eradication of untouchability, progressive amelioration of the toiling people, social enfranchisement of women, free and compulsory primary education and overhauling of the system of higher education so as to meet the requirements of the ordinary people instead of the middle class. It is striking to note that most of these issues form integral part of Agenda-21 of the Rio Summit which gave a blue print for sustainable development.

One of the defining features of modern civilization is the annihilation of distance by excessively using motorized transport. Proliferation of cars and air planes to make communication easier for enhancing mobility and making the world smaller have choked peoples across the globe with air pollution and emission of green house gases. Joseph Stizlitz in his book 'Making Globalisation Work' has written that while 80% of the global warming is caused by hydrocarbons and 20% is caused by deforestation. Increasingly, more and more people are possessing cars which are symbols of status, individuality and mobility. The threat posed by growing number of cars to environment is well known. Now it is being asked if planet earth can cope up with the toxic emissions from 4 billion cars possessed by peoples in America, China, India and Europe. The ability of people to have cars and provide fuel to them does not augur well for the climate. Combined with refrigerator and air-conditioning it will cause irreversible damage to the ozone layer and carrying capacity of the earth. Annihilation of distance coupled with pursuit of comfort will further contribute to unsustainable development.

The craze for car began in 1930s when the President of America Mr. Hoover outlined his plan for two cars and two radio sets for each American family. Mahatma Gandhi was informed about it by an American correspondent and requested to outline his future vision of Indian society. He in his characteristic way replied that if every Indian family would possess a car there would be so many of them resulting in lack of space to walk. Adding further he stated that in his vision of Indian society possession of a car would not be considered a meritorious thing.

Again during the Dandi March when some people brought oranges in a motorized transport he disapproved of it and said, "The rule should be avoid the car if you can walk." There are many European countries where congestion tax is imposed for cars to enter certain key areas to keep them free from vehicular pollution. There are several other countries in Europe which are adopting a car less day. They have realized the demerit of possessing too many cars. In other words the utterances of Mahatma Gandhi concerning cars are being realized with added poignancy.

Going beyond the terrestrial sphere we find that the civil aviation sector is growing in an unprecedented scale and thriving by introducing cheaper fares for passengers. It is contributing to the greater integration of different parts of the country and world. Annihilation of distance through air planes is not an unmixed blessing. The London Economist in its issue of 10th June, 2006 carried a cover story under the caption "The Dirty Sky: Why Air Travel will be the Next Green Battle Ground". It observed "Put frankly, air travel makes a mockery of many peoples attempts to live a green life. Somebody who wants to reduce his "carbon footprint" can bicycle to work, never buy aerosols and turn off his air conditioner – and still blow away all this virtue on a couple of long flights."

Stating that "Air transport is the fastest growing source of green house gas emission but so far sparked relatively little concern among Governments and international bodies" she wrote "One return flight to, say, Miami, and you are responsible for more carbon dioxide production than a year's motoring".

These grave concerns expressed in foreign press starkly remind us about Mahatma Gandhi's reservations about annihilation of distance. In Hind Swaraj, he described railways a necessary evil. All other faster means of communication can indeed constitute necessary evil. The necessary evil stretched beyond a point will overwhelm mankind and the planet earth. It is in this context that his wise counsel not to subordinate human interest to machine assumes paramount significance.

While dealing with Mahatma Gandhi and Sustainable Development one would inevitably deal with the question of poverty which is the worst source of pollution. Fight for eradication of poverty by using appropriate technology and non-violent means is nothing but a fight for sustainable development. Mahatma Gandhi wrote Hind Swaraj not only to criticize modern civilization but also to eradicate poverty in India. Smt. Indira Gandhi, former Prime Minister of India in her speech in the first ever U.N. Conference on Environment organised in Stockholm in 1972 declared that poverty is the worst source of pollution. Mahatma Gandhi by addressing the issue of eradication of poverty in Hind Swaraj was addressing the core issue of sustainable development. Almost eight decades after Mahatma Gandhi did deal with the issue of poverty. The Brundtland Commission on "Our Common Future" wrote "A world in which poverty is endemic will always be prone to ecological and other catastrophes".

Adoption of a comparatively simple life style by the western people will be a major factor for eradication of poverty. In fact the life style of the peoples of the advanced countries have gravely endangered environment. They have given priority to their own interests over the interests of nature. Prof. David H. Bennett wrote "The Cinderella Syndrome: Ecologically Sustainable Development and Ecological Competence – A Second Precautionary Tale" wrote that aborigines in Australia during 40,000 to 80,000 years have done much less damage to the continent than the non-aborigines inhabitants in the last 200 years. He holds that the technological imperative of the non-aborigine inhabitant contributed to the damage of the ecology and exhorts them to learn the lessons of restraint and ecological competence from the aborigines to adopt a sustainable way of life. At the end he captures the spirit of Mahatma Gandhi by writing "Dominant western cultures must learn to live simply so that others can simply live." It is this simple living which can go a long way in addressing the issue of poverty eradication and promoting the cause of sustainable development.

The discussion of Mahatma Gandhi and Sustainable Development would be incomplete without referring to the burning issue of water scarcity in the world. Twenty first century has been described as the most water stressed century in the world. Water famines across the world may cause conflicts among nations. If not controlled and dealt with in a fair and equitable manner the water scarcity problem may give rise to another world war reminiscent of other world wars over resources and other trading and commercial interests. It is in this context that Mahatma Gandhi's ideas need to be recollected and put into practice.

During our struggle for independence he referred to the water famine occurring in the Kathiawar region of Gujarat ruled by many princes. To address the issue of acute shortage of water he advised all the princely States to form a confederation and take long term measures for planting trees in vast tracts of land. He opined that afforestation on a large scale constituted the most effective step to face the water crisis. The twenty-first century world need to follow his words with utmost seriousness. The British rulers who treated forests as a source of revenue hardly understood their relevance from the point of view of ecology and sustainable development. Their approach was a byproduct of the exploitation of natural resources regardless of its consequences for the common people and environment.

Tuning himself with the common people whom he called "the dumb millions" he also suggested in a prayer meeting in Delhi in 1947 for harvesting rain water and using it for irrigational purposes to avoid famines and food shortages. The M.S. Swaminathan Commission for Farmers in its report submitted to the Prime Minister in 2006 recommended to harvest rain water for addressing the problem of irrigation affecting our farmers.

Mahatma Gandhi was far ahead of his times in grappling with challenges to planet earth arising out of a life style which multiplied wants and desires and left no stone unturned to satisfy them. At a time when mankind is facing the dangerous prospects of getting annihilated due to accelerating pace of global warming it is important to rediscover Gandhiji's ideas and put them into practice. It is heartening that in many parts of the world people are getting inspired by his ideals and taking appropriate action. It was best reflected in the initiative taken in Germany to establish Green Party and pursue policies consistent with nature and ecology. One of the founders of Green Party Ms. Patra Kelly admirably summed up the impact of Mahatma Gandhi in forming the party when she wrote the following:

> In particular area of our work we have been greatly inspired by Mahatma Gandhi. That is in our belief that a life style and method of production which rely on endless supply of raw materials and which use those raw materials lavishly, also provide motive force for violent appropriation of raw materials from other countries. In contrast a responsible use of raw materials as part of an ecologically oriented life style and economy reduces the risk that policies of violence will pursue.

Such a vision provides the remedy to create a new civilization the foundation of which is based on discipline, restraint and morality. It is heartening to note that the recent literature being brought out in the western world is eloquently following the vision of Mahatma Gandhi. A book "Surviving the Century: Facing Climate Chaos" stresses on measures suggested by Mahatma Gandhi in the beginning of the twentieth century. The book argues for an approach which would speak for the earth community. It suggests that such an approach can be devised if we become non-violent, respect nature, follow the path of sustainable development and ensure justice to the poor. All those aspects remained central to Mahatma Gandhi's life and work. There is slow but sure realization that by following Gandhiji's ideals we can survive the century. The line of argument which attempts to speak for the earth community essentially recaptures the immortal and eloquent words of Mahatma Gandhi that earth has enough for fulfilling everybody's need but not anybody's greed. These words constitute the sum and substance of sustainable development. There is no alternative to such a world view. The Time Magazine in its 9th April 2007 issue came out with 51 Global Warming Survival Guides. The 51st Guide earnestly suggests to share more, consume less and simplify life.In other words the Time Magazine, one of the mouth pieces of the western world, is turning to Mahatma Gandhi to save the world from the danger of extinction caused by global warming. It is a measure of Mahatma Gandhi's enduring and deeper significance in the context of attempts to protect the planet earth. It is therefore indispensable to rediscover his writings and comprehend them to further the cause of sustainable development.

Strategies for Sustainable Development

4

There are various kinds of strategies used for sustainable development. A few of them are sustainable agriculture, sustainable transport and sustainable forest management. This chapter closely examines these key strategies to provide an extensive understanding of the subject.

The Rio Summit established sustainable development as the guiding vision for the development efforts of all countries. The strategies for sustainable development to ensure socially responsible economic development while protecting the resource base and the environment for the benefit of future generations. To agree how the international community can best assist developing countries in meeting good of sustainability.

Commitment to provide support for sound nationally-owned sustainable development strategies where conditions for effective partnership are in place. In simple terms, sustainable development means integrating the economic, social and environmental objectives of society, in order to maximize human well-being in the present without compromising the ability of future generations to meet their needs.

Strategy Formulation

- Country ownership and participation, leadership and initiative in developing their strategies.

- Broad consultation, including particularly with the poor and with civil society, to open up debate on new ideas and information, expose issues to be addressed, and build consensus and political support on action.

- Ensuring sustained beneficial impacts on disadvantaged and marginalized groups and on future generations.

- Building on existing strategies and processes, rather than adding additional ones, to enable convergence and coherence.

- A solid analytical basis, taking account also of relevant regional issues, including a comprehensive review of the present situation and forecasts of trends and risks.

- Integration of economic, social and environmental objectives.

- Through mutually supportive policies and practices and the management of tradeoffs.

- Realistic targets with clear budgetary priorities.

The environment is a key determinant of growth and of poverty reduction. Environmental issues, including longer-term and global perspectives, need to be integrated into mainstream planning processes affecting these and other development objectives.

SUSTAINABLE AGRICULTURE

Sustainable architecture is architecture that seeks to minimize the negative environmental impact of buildings by efficiency and moderation in the use of materials, energy, and development space and the ecosystem at large. Sustainable architecture uses a conscious approach to energy and ecological conservation in the design of the built environment.

Hanging gardens of One Central Park, Sydney.

Energy-plus-houses at Freiburg-Vauban in Germany.

The idea of sustainability, or ecological design, is to ensure that our use of presently available resources does not end up having detrimental effects to our collective well-being or making it impossible to obtain resources for other applications in the long run.

Sustainable Energy Use

Energy efficiency over the entire life cycle of a building is the most important goal of sustainable architecture. Architects use many different passive and active techniques to reduce the energy needs of buildings and increase their ability to capture or generate their own energy. One of the keys to exploit local environmental resources and influence energy-related factors such as daylight, solar heat gains and ventilation is the use of site analysis.

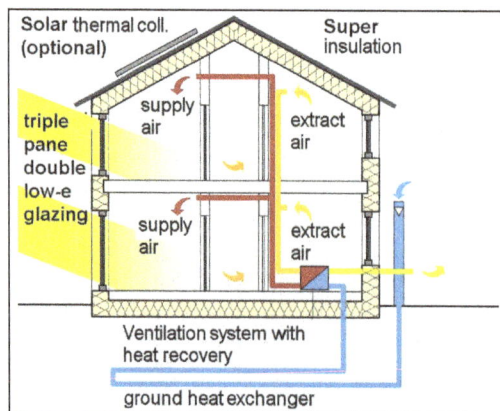

The passivhaus standard combines a variety of techniques and technologies to achieve ultra-low energy use.

K2 sustainable apartments in Windsor, Victoria, Australia features passive solar design, recycled and sustainable materials, photovoltaic cells, wastewater treatment, rainwater collection and solar hot water.

The town's new art center, which integrates its own solar panels and wind generators for energy self-sufficiency.

Heating, Ventilation and Cooling System Efficiency

Numerous passive architectural strategies have been developed over time. Examples of such strategies include the arrangement of rooms or the sizing and orientation of windows in a building, and the orientation of facades and streets or the ratio between building heights and street widths for urban planning.

An important and cost-effective element of an efficient heating, ventilating, and air conditioning (HVAC) system is a well-insulated building. A more efficient building requires less heat generating or dissipating power, but may require more ventilation capacity to expel polluted indoor air.

Significant amounts of energy are flushed out of buildings in the water, air and compost streams. Off the shelf, on-site energy recycling technologies can effectively recapture energy from waste hot water and stale air and transfer that energy into incoming fresh cold water or fresh air. Recapture of energy for uses other than gardening from compost leaving buildings requires centralized anaerobic digesters.

HVAC systems are powered by motors. Copper, versus other metal conductors, helps to improve the electrical energy efficiencies of motors, thereby enhancing the sustainability of electrical building components. Site and building orientation have some major effects on a building's HVAC efficiency.

Passive solar building design allows buildings to harness the energy of the sun efficiently without the use of any active solar mechanisms such as photovoltaic cells or solar hot water panels. Typically passive solar building designs incorporate materials with high thermal mass that retain heat effectively and strong insulation that works to prevent heat escape. Low energy designs also requires the use of solar shading, by means of awnings, blinds or shutters, to relieve the solar heat gain in summer and to reduce the need for artificial cooling. In addition, low energy buildings typically have a very low surface area to volume ratio to minimize heat loss. This means that sprawling multi-winged building designs (often thought to look more "organic") are often avoided in favor of more centralized structures. Traditional cold climate buildings such as American colonial saltbox designs provide a good historical model for centralized heat efficiency in a small-scale building.

Windows are placed to maximize the input of heat-creating light while minimizing the loss of heat through glass, a poor insulator. In the northern hemisphere this usually involves installing a large number of south-facing windows to collect direct sun and severely restricting the number of north-facing windows. Certain window types, such as double or triple glazed insulated windows with gas filled spaces and low emissivity (low-E) coatings, provide much better insulation than single-pane glass windows. Preventing excess solar gain by means of solar shading devices in the summer months is important to reduce cooling needs. Deciduous trees are often planted in front of windows to block excessive sun in summer with their leaves but allow light through in winter when their leaves fall off. Louvers or light shelves are installed to allow the sunlight in during the winter (when the sun is lower in the sky) and keep it out in the summer (when the sun is high in the sky). Coniferous or evergreen plants are often planted to the north of buildings to shield against cold north winds.

In colder climates, heating systems are a primary focus for sustainable architecture because they are typically one of the largest single energy drains in buildings.

In warmer climates where cooling is a primary concern, passive solar designs can also be very effective. Masonry building materials with high thermal mass are very valuable for retaining the cool temperatures of night throughout the day. In addition builders often opt for sprawling single story structures in order to maximize surface area and heat loss. Buildings are often designed to capture and channel existing winds, particularly the especially cool winds coming from nearby bodies of water. Many of these valuable strategies are employed in some way by the traditional architecture of warm regions, such as south-western mission buildings.

In climates with four seasons, an integrated energy system will increase in efficiency: when the building is well insulated, when it is sited to work with the forces of nature, when heat is recaptured (to be used immediately or stored), when the heat plant relying on fossil fuels or electricity is greater than 100% efficient, and when renewable energy is used.

Renewable Energy Generation

BedZED (Beddington Zero Energy Development), the UK's largest and first carbon-neutral eco-community: the distinctive roofscape with solar panels and passive ventilation chimneys.

Solar Panels

Active solar devices such as photovoltaic solar panels help to provide sustainable electricity for any use. Electrical output of a solar panel is dependent on orientation, efficiency, latitude, and climate—solar gain varies even at the same latitude. Typical efficiencies for commercially available PV panels range from 4% to 28%. The low efficiency of certain photovoltaic panels can significantly affect the payback period of their installation. This low efficiency does not mean that solar panels are not a viable energy alternative. In Germany for example, Solar Panels are commonly installed in residential home construction.

Roofs are often angled toward the sun to allow photovoltaic panels to collect at maximum efficiency. In the northern hemisphere, a true-south facing orientation maximizes yield for solar panels. If true-south is not possible, solar panels can produce adequate energy if aligned within 30° of south. However, at higher latitudes, winter energy yield will be significantly reduced for non-south orientation.

To maximize efficiency in winter, the collector can be angled above horizontal Latitude +15°. To maximize efficiency in summer, the angle should be Latitude -15°. However, for an annual maximum production, the angle of the panel above horizontal should be equal to its latitude.

Wind Turbines

The use of undersized wind turbines in energy production in sustainable structures requires the consideration of many factors. In considering costs, small wind systems are generally more expensive than larger wind turbines relative to the amount of energy they produce. For small wind turbines, maintenance costs can be a deciding factor at sites with marginal wind-harnessing capabilities. At low-wind sites, maintenance can consume much of a small wind turbine's revenue. Wind turbines begin operating when winds reach 8 mph, achieve energy production capacity at speeds of 32-37 mph, and shut off to avoid damage at speeds exceeding 55 mph. The energy potential of a wind turbine is proportional to the square of the length of its blades and to the cube of the speed at which its blades spin. Though wind turbines are available that can supplement power for a single building, because of these factors, the efficiency of the wind turbine depends much upon the wind conditions at the building site. For these reasons, for wind turbines to be at all efficient, they must be installed at locations that are known to receive a constant amount of wind (with average wind speeds of more than 15 mph), rather than locations that receive wind sporadically. A small wind turbine can be installed on a roof. Installation issues then include the strength of the roof, vibration, and the turbulence caused by the roof ledge. Small-scale rooftop wind turbines have been known to be able to generate power from 10% to up to 25% of the electricity required of a regular domestic household dwelling. Turbines for residential scale use are usually between 7 feet (2 m) to 25 feet (8 m) in diameter and produce electricity at a rate of 900 watts to 10,000 watts at their tested wind speed. Building integrated wind turbine performance can be enhanced with the addition of an aerofoil wing on top of a roof mounted turbine.

Solar Water Heating

Solar water heaters, also called solar domestic hot water systems, can be a cost-effective way to generate hot water for a home. They can be used in any climate, and the fuel they use—sunshine—is free.

There are two types of solar water systems- active and passive. An active solar collector system can produce about 80 to 100 gallons of hot water per day. A passive system will have a lower capacity.

There are also two types of circulation, direct circulation systems and indirect circulation systems. Direct circulation systems loop the domestic water through the panels. They should not be used in climates with temperatures below freezing. Indirect circulation loops glycol or some other fluid through the solar panels and uses a heat exchanger to heat up the domestic water.

The two most common types of collector panels are Flat-Plate and Evacuated-tube. The two work similarly except that evacuated tubes do not convectively lose heat, which greatly improves their efficiency (5%-25% more efficient). With these higher efficiencies, Evacuated-tube solar collectors can also produce higher-temperature space heating, and even higher temperatures for absorption cooling systems.

Electric-resistance water heaters that are common in homes today have an electrical demand around 4500 kWh/year. With the use of solar collectors, the energy use is cut in half. The up-front cost of installing solar collectors is high, but with the annual energy savings, payback periods are relatively short.

Heat Pumps

Air-source heat pumps (ASHP) can be thought of as reversible air conditioners. Like an air conditioner, an ASHP can take heat from a relatively cool space (e.g. a house at 70 °F) and dump it into a hot place (e.g. outside at 85 °F). However, unlike an air conditioner, the condenser and evaporator of an ASHP can switch roles and absorb heat from the cool outside air and dump it into a warm house.

Air-source heat pumps are inexpensive relative to other heat pump systems. However, the efficiency of air-source heat pumps decline when the outdoor temperature is very cold or very hot; therefore, they are only really applicable in temperate climates.

For areas not located in temperate climates, ground-source (or geothermal) heat pumps provide an efficient alternative. The difference between the two heat pumps is that the ground-source has one of its heat exchangers placed underground—usually in a horizontal or vertical arrangement. Ground-source takes advantage of the relatively constant, mild temperatures underground, which means their efficiencies can be much greater than that of an air-source heat pump. The in-ground heat exchanger generally needs a considerable amount of area. Designers have placed them in an open area next to the building or underneath a parking lot.

Energy Star ground-source heat pumps can be 40% to 60% more efficient than their air-source counterparts. They are also quieter and can also be applied to other functions like domestic hot water heating.

In terms of initial cost, the ground-source heat pump system costs about twice as much as a standard air-source heat pump to be installed. However, the up-front costs can be more than offset by the decrease in energy costs. The reduction in energy costs is especially apparent in areas with typically hot summers and cold winters.

Other types of heat pumps are water-source and air-earth. If the building is located near a body of water, the pond or lake could be used as a heat source or sink. Air-earth heat pumps circulate the building's air through underground ducts. With higher fan power requirements and inefficient heat transfer, air-earth heat pumps are generally not practical for major construction.

Sustainable Building Materials

Some examples of sustainable building materials include recycled denim or blown-in fiber glass insulation, sustainably harvested wood, Trass, Linoleum, sheep wool, concrete (high and ultra high performance roman self-healing concrete), panels made from paper flakes, baked earth, rammed earth, clay, vermiculite, flax linnen, sisal, seegrass, expanded clay grains, coconut, wood fiber plates, calcium sandstone, locally obtained stone and rock, and bamboo, which is one of the strongest and fastest growing woody plants, and non-toxic low-VOC glues and paints. Vegetative cover or shield over building envelopes also helps in the same. Paper which is fabricated or manufactured out of forest wood is supposedly hundred percent recyclable. Thus it regenerates and saves almost all the forest wood that it takes during its manufacturing process.

Recycled Materials

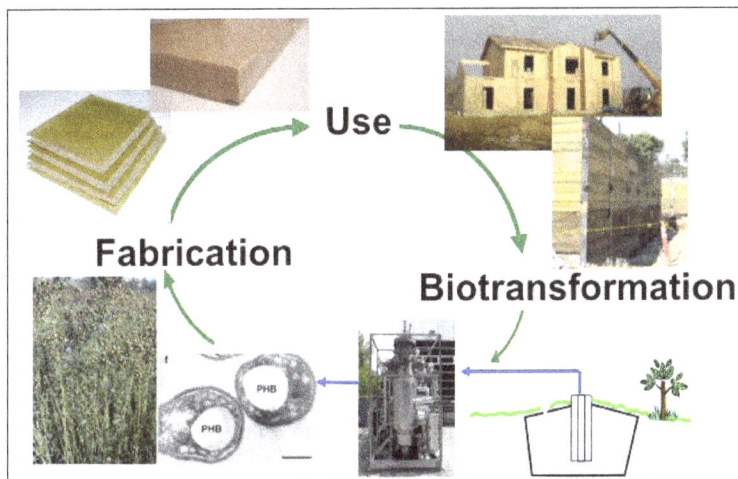

Recycling items for building.

Sustainable architecture often incorporates the use of recycled or second hand materials, such as reclaimed lumber and recycled copper. The reduction in use of new materials creates a corresponding reduction in embodied energy (energy used in the production of materials). Often sustainable architects attempt to retrofit old structures to serve new needs in order to avoid unnecessary development. Architectural salvage and reclaimed materials are used when appropriate. When older buildings are demolished, frequently any good wood is reclaimed, renewed, and sold as flooring. Any good dimension stone is similarly reclaimed. Many other parts are reused as well, such as doors, windows, mantels, and hardware, thus reducing the consumption of new goods. When new materials are employed, green designers look for materials that are rapidly replenished, such as bamboo, which can be harvested for commercial use after only 6 years of growth, sorghum or wheat straw, both of which are waste material that can be pressed into panels, or cork oak, in which only the outer bark is removed for use, thus preserving the tree. When possible, building materials may be gleaned from the site itself; for example, if a new structure is being constructed in a wooded area, wood from the trees which were cut to make room for the building would be re-used as part of the building itself.

Lower Volatile Organic Compounds

Low-impact building materials are used wherever feasible: for example, insulation may be made from low VOC (volatile organic compound)-emitting materials such as recycled denim or cellulose insulation, rather than the building insulation materials that may contain carcinogenic or toxic materials such as formaldehyde. To discourage insect damage, these alternate insulation materials may be treated with boric acid. Organic or milk-based paints may be used. However, a common fallacy is that "green" materials are always better for the health of occupants or the environment. Many harmful substances (including formaldehyde, arsenic, and asbestos) are naturally occurring and are not without their histories of use with the best of intentions. A study of emissions from materials by the State of California has shown that there are some green materials that have substantial emissions whereas some more "traditional" materials actually were lower emitters. Thus, the subject of emissions must be carefully investigated before concluding that natural materials are always the healthiest alternatives for occupants and for the Earth.

Volatile organic compounds (VOC) can be found in any indoor environment coming from a variety of different sources. VOCs have a high vapor pressure and low water solubility, and are suspected of causing sick building syndrome type symptoms. This is because many VOCs have been known to cause sensory irritation and central nervous system symptoms characteristic to sick building syndrome, indoor concentrations of VOCs are higher than in the outdoor atmosphere, and when there are many VOCs present, they can cause additive and multiplicative effects.

Green products are usually considered to contain fewer VOCs and be better for human and environmental health. A study conducted by the Department of Civil, Architectural, and Environmental Engineering at the University of Miami that compared three green products and their non-green counterparts found that even though both the green products and the non-green counterparts both emitted levels of VOCs, the amount and intensity of the VOCs emitted from the green products were much safer and comfortable for human exposure.

Materials Sustainability Standards

Despite the importance of materials to overall building sustainability, quantifying and evaluating the sustainability of building materials has proven difficult. There is little coherence in the measurement and assessment of materials sustainability attributes, resulting in a landscape today that is littered with hundreds of competing, inconsistent and often imprecise eco-labels, standards and certifications. This discord has led both to confusion among consumers and commercial purchasers and to the incorporation of inconsistent sustainability criteria in larger building certification programs such as LEED. Various proposals have been made regarding rationalization of the standardization landscape for sustainable building materials.

Waste Management

Waste takes the form of spent or useless materials generated from households and businesses, construction and demolition processes, and manufacturing and agricultural industries. These materials are loosely categorized as municipal solid waste, construction and demolition (C&D) debris, and industrial or agricultural by-products. Sustainable architecture focuses on the on-site use of waste management, incorporating things such as grey water systems for use on garden beds, and composting toilets to reduce sewage. These methods, when combined with on-site food waste composting and off-site recycling, can reduce a house's waste to a small amount of packaging waste.

Building Placement

One central and often ignored aspect of sustainable architecture is building placement. Although the ideal environmental home or office structure is often envisioned as an isolated place, this kind of placement is usually detrimental to the environment. First, such structures often serve as the unknowing frontlines of suburban sprawl. Second, they usually increase the energy consumption required for transportation and lead to unnecessary auto emissions. Ideally, most building should avoid suburban sprawl in favor of the kind of light urban development articulated by the New Urbanist movement. Careful mixed use zoning can make commercial, residential, and light industrial areas more accessible for those traveling by foot, bicycle, or public transit, as proposed in the Principles of Intelligent Urbanism. The study of Permaculture, in its holistic application, can also greatly help in proper building placement that minimizes energy consumption and works with the surroundings rather than against them, especially in rural and forested zones.

Sustainable Building Consulting

A sustainable building consultant may be engaged early in the design process, to forecast the sustainability implications of building materials, orientation, glazing and other physical factors, so as to identify a sustainable approach that meets the specific requirements of a project.

Norms and standards have been formalized by performance-based rating systems e.g. LEED and Energy Star for homes. They define benchmarks to be met and provide metrics and testing to meet those benchmarks. It is up to the parties involved in the project to determine the best approach to meet those standards.

Changing Pedagogues

Critics of the reductionism of modernism often noted the abandonment of the teaching of architectural history as a causal factor. The fact that a number of the major players in the shift away from modernism were trained at Princeton University's School of Architecture, where recourse to history continued to be a part of design training in the 1940s and 1950s, was significant. The increasing rise of interest in history had a profound impact on architectural education. History courses became more typical and regularized. With the demand for professors knowledgeable in the history of architecture, several PhD programs in schools of architecture arose in order to differentiate themselves from art history PhD programs, where architectural historians had previously trained. In the US, MIT and Cornell were the first, created in the mid-1970s, followed by Columbia, Berkeley, and Princeton. Among the founders of new architectural history programs were Bruno Zevi at the Institute for the History of Architecture in Venice, Stanford Anderson and Henry Millon at MIT, Alexander Tzonis at the Architectural Association, Anthony Vidler at Princeton,

Manfredo Tafuri at the University of Venice, Kenneth Frampton at Columbia University, and Werner Oechslin and Kurt Forster at ETH Zürich.

The term "sustainability" in relation to architecture has so far been mostly considered through the lens of building technology and its transformations. Going beyond the technical sphere of "green" design, invention and expertise, some scholars are starting to position architecture within a much broader cultural framework of the human interrelationship with nature. Adopting this framework allows tracing a rich history of cultural debates about our relationship to nature and the environment, from the point of view of different historical and geographical contexts.

Sustainable Urbanism and Architecture

Concurrently, the recent movements of New Urbanism and New Classical Architecture promote a sustainable approach towards construction, that appreciates and develops smart growth, architectural tradition and classical design. This in contrast to modernist and globally uniform architecture, as well as leaning against solitary housing estates and suburban sprawl. Both trends started in the 1980s. The Driehaus Architecture Prize is an award that recognizes efforts in New Urbanism and New Classical Architecture, and is endowed with a prize money twice as high as that of the modernist Pritzker Prize.

Building Information Modelling

Building Information Modelling (BIM) is used to help enable sustainable design by allowing architects and engineers to integrate and analyze building performance. BIM services, including conceptual and topographic modelling, offer a new channel to green building with successive and immediate availability of internally coherent, and trustworthy project information. BIM enables designers to quantify the environmental impacts of systems and materials to support the decisions needed to design sustainable buildings.

ECO-INDUSTRIAL DEVELOPMENT

Eco-industrial development (EID) is a framework for industry to develop while reducing its impact on the environment. It uses a closed loop production cycle to tackle a broad set of environmental challenges such as soil and water pollution, desertification, species preservation, energy management, by-product synergy, resource efficiency, air quality, etc.

Mutually beneficial connections among industry, natural systems, energy, material and local communities become central factors in designing industrial production processes.

The approach itself is largely voluntary and market-driven but often pressed ahead by favorable government treatment or efforts of development co-operation.

Since the early 1990s the idea of EID evolved from biological symbiosis. This concept was adapted by industrial ecologists in the search for innovative approaches to solve problems of waste, energy shortage and degradation of the environment. A continuous approach towards improving both environmental and economic outcomes is used.

In 1992, the international community officially connected development co-operation to sustainable environmental protection for the first time. At the United Nations Conference on Environment and Development (UNCED) in Rio de Janeiro, Brazil nearly 180 states signed the conference's Rio Declaration. Although non-binding, the Rio Declaration on Environment and Development laid out 27 principles that shall guide the increasing inter-connectedness of development cooperation and sustainability. Moreover, the documents drafting was accompanied by a presentation describing the idea of eco-industrial development for the first time.

In the following years, EID became popular throughout the United States. The recently elected Clinton administration installed a summit of business, labor, government and environmental protection representatives to further develop the approach. This summit established the idea of eco-industrial parks but soon realized that at first a more efficient management of raw materials, energy and waste has to be achieved.

Since then, the broad goals and application principles of EID have hardly changed and only became adapted to the rapid technological progress.

Goals and Concepts

The primary goal of eco-industrial development is a significant and continuous improvement in both business and environmental performance. Herein, the notion of industry not only relates to private-sector manufacturing but also covers state-owned enterprise, the service sector as well as transportation. EID's twin guideline is reflected specifically in the "eco" of eco-industrial as it resembles ecology (decrease in pollution and waste) and economy (increase in commercial success) at the same time. In order to build a framework of defining an enterprise's sustainable performance at the micro level, resource use optimization, minimization of waste, cleaner technologies and pollution limits are used in achieving a broad range of goals in EID:

- Resource efficiency minimizes the use of energy, materials, water and transportation. This, in turn, lowers production costs due to savings in virtually all areas of business.

- Cleaner production is a predominantly environmental measure, which aims at the reduction or even substitution of toxics, emissions-control or the re-use of residual material.

- Renewables in both energy and material use shall eliminate all pollution through fossil fuels.

- Greening of buildings or production sites installs high energy and environmental standards by relying on innovation in green architecture or engineering. Moreover, new facility and infrastructure design may also enhance the quality of life in neighboring communities significantly.

- Environmental management systems such as the ISO 14000 ensure a continuous improvement through regular audits and the progressing establishment of environmental targets.

- Ecological site planning can then combine each of these aspects by developing a clear understanding of air, water and ground system capacities throughout the surrounding eco-system.

Eco-industrial development hence explores the possibility of improvement at the local level. In unique case-to-case analyses, particular geography, human potential or business climate are investigated. In contrast to the widespread race for top-down governmental support such as tax cuts, EID emphasizes locally achievable success and rooms for improvement. As a result, purposeful enforcements of action plans can make a large difference by optimizing the interaction of business, community and ecological systems.

Instruments

Eco-industrial development includes and employs four major conceptual instruments. Each of the approaches intends to combine the seemingly antithetic processes of industrial development and bolstering sustainability.

- Industrial Ecology focuses on both industrial as well as consumer behavior. By assessing flows of energy and material, the approach determines the flows influences on the environment. In turn, it explores ways and means of optimizing the whole production chain from flow and use of resources to their final transformation. During these analyses, influences of economic, political, regulatory and social factors are key.

- The concept of Industrial Symbiosis is based on mainly voluntary cooperation of different industries. By conglomerating complementary enterprises and by then adapting their respective production chains, the presence of each may increase viability and profitability of the others. Therefore, symbioses consider resource scarcity and environmental protection as crucial factors in developing sustainable industries and profits. Industrial Symbiosis often becomes manifest in Eco-industrial parks.

- Environmental Management Systems include a wide range of different environmental management approaches in order to ensure continual improvement in sustainability. In early stages, monitoring companies facilitates the identification of hazardous environmental aspects. Further on, objectives and targets are set under consideration of legal requirements. Finally, the establishment of regular audits and other reporting systems combined with continuous follow-up targets shall ensure a constant improvement towards greener industrial production.

- The Design for the Environment concept originated in engineering disciplines as well as from the product life-cycle analysis. It is a simple but all-encompassing assessment of a product's potential environmental impact – ranging from energy and materials used for packaging, transportation, consumer use and disposal.

SUSTAINABLE TRANSPORT

Sustainable transport refers to the broad subject of transport that is sustainable in the senses of social, environmental and climate impacts. Components for evaluating sustainability include the particular vehicles used for road, water or air transport; the source of energy; and the infrastructure used to accommodate the transport (roads, railways, airways, waterways, canals and terminals). Transport operations and logistics as well as transit-oriented development are also involved in evaluation. Transportation sustainability is largely being measured by transportation system effectiveness and efficiency as well as the environmental and climate impacts of the system.

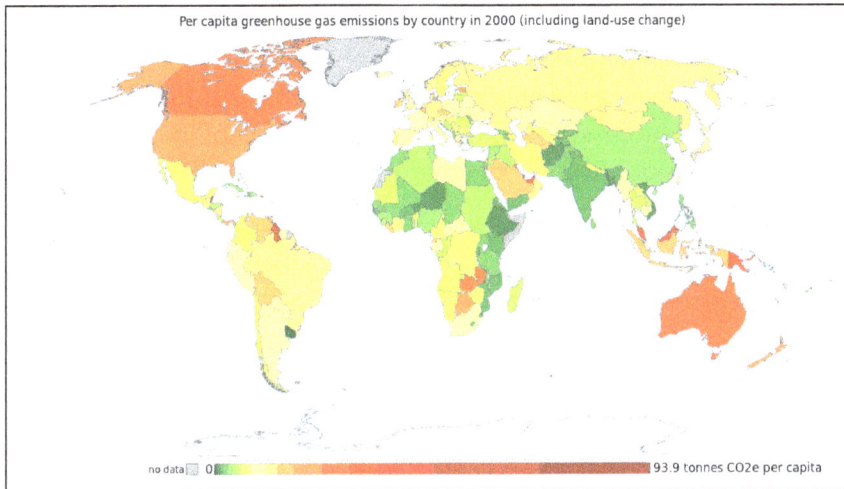

Anthropogenic per capita emissions of greenhouse gases by country by the year 2000.

Short-term activity often promotes incremental improvement in fuel efficiency and vehicle emissions controls while long-term goals include migrating transportation from fossil-based energy to other alternatives such as renewable energy and use of other renewable resources. The entire life cycle of transport systems is subject to sustainability measurement and optimization.

Sustainable transport systems make a positive contribution to the environmental, social and economic sustainability of the communities they serve. Transport systems exist to provide social and economic connections, and people quickly take up the opportunities offered by increased mobility, with poor households benefiting greatly from low carbon transport options. The advantages of increased mobility need to be weighed against the environmental, social and economic costs that transport systems pose.

Transport systems have significant impacts on the environment, accounting for between 20% and 25% of world energy consumption and carbon dioxide emissions. The majority of the emissions, almost 97%, came from direct burning of fossil fuels. Greenhouse gas emissions from transport are increasing at a faster rate than any other energy using sector. Road transport is also a major contributor to local air pollution and smog.

The United Nations Environment Programme (UNEP) estimates that each year 2.4 million premature deaths from outdoor air pollution could be avoided. Particularly hazardous for health are emissions of black carbon, a component of particulate matter, which is a known cause of respiratory and carcinogenic diseases and a significant contributor to global climate change. The links between greenhouse gas emissions and particulate matter make low carbon transport an increasingly sustainable investment at local level—both by reducing emission levels and thus mitigating climate change; and by improving public health through better air quality.

The social costs of transport include road crashes, air pollution, physical inactivity, time taken away from the family while commuting and vulnerability to fuel price increases. Many of these negative impacts fall disproportionately on those social groups who are also least likely to own and drive cars. Traffic congestion imposes economic costs by wasting people's time and by slowing the delivery of goods and services.

Traditional transport planning aims to improve mobility, especially for vehicles, and may fail to adequately consider wider impacts. But the real purpose of transport is access – to work, education, goods and services, friends and family – and there are proven techniques to improve access while simultaneously reducing environmental and social impacts, and managing traffic congestion. Communities which are successfully improving the sustainability of their transport networks are doing so as part of a wider programme of creating more vibrant, livable, sustainable cities.

The term sustainable transport came into use as a logical follow-on from sustainable development, and is used to describe modes of transport, and systems of transport planning, which are consistent with wider concerns of sustainability. There are many definitions of the sustainable transport, and of the related terms sustainable transportation and sustainable mobility. One such definition, from the European Union Council of Ministers of Transport, defines a sustainable transportation system as one that:

- Allows the basic access and development needs of individuals, companies and society to be met safely and in a manner consistent with human and ecosystem health, and promotes equity within and between successive generations.

- Is affordable, operates fairly and efficiently, offers a choice of transport mode, and supports a competitive economy, as well as balanced regional development.

- Limits emissions and waste within the planet's ability to absorb them, uses renewable resources at or below their rates of generation, and uses non-renewable resources at or below the rates of development of renewable substitutes, while minimizing the impact on the use of land and the generation of noise.

Sustainability extends beyond just the operating efficiency and emissions. A life-cycle assessment involves production, use and post-use considerations. A cradle-to-cradle design is more important than a focus on a single factor such as energy efficiency.

Most of the tools and concepts of sustainable transport were developed before the phrase was coined. Walking, the first mode of transport, is also the most sustainable. Public transport dates back at least as far as the invention of the public bus by Blaise Pascal in 1662. The first passenger tram began operation in 1807 and the first passenger rail service in 1825. Pedal bicycles date from the 1860s. These were the only personal transport choices available to most people in Western countries prior to World War II, and remain the only options for most people in the developing world. Freight was moved by human power, animal power or rail.

The post-war years brought increased wealth and a demand for much greater mobility for people and goods. The number of road vehicles in Britain increased fivefold between 1950 and 1979, with similar trends in other Western nations. Most affluent countries and cities invested heavily in bigger and better-designed roads and motorways, which were considered essential to underpin growth and prosperity. Transport planning became a branch of Urban Planning and identified induced demand as a pivotal change from "predict and provide" toward a sustainable approach incorporating land use planning and public transit. Public investment in transit, walking and cycling declined dramatically in the United States, Great Britain and Australia, although this did not occur to the same extent in Canada or mainland Europe.

Concerns about the sustainability of this approach became widespread during the 1973 oil crisis and the 1979 energy crisis. The high cost and limited availability of fuel led to a resurgence of interest in alternatives to single occupancy vehicle travel.

Transport innovations dating from this period include high-occupancy vehicle lanes, citywide carpool systems and transportation demand management. Singapore implemented congestion pricing in the late 1970s, and Curitiba began implementing its Bus Rapid Transit system in the early 1980s.

Relatively low and stable oil prices during the 1980s and 1990s led to significant increases in vehicle travel from 1980–2000, both directly because people chose to travel by car more often and for greater distances, and indirectly because cities developed tracts of suburban housing, distant from shops and from workplaces, now referred to as urban sprawl. Trends in freight logistics, including a movement from rail and coastal shipping to road freight and a requirement for just in time deliveries, meant that freight traffic grew faster than general vehicle traffic.

At the same time, the academic foundations of the "predict and provide" approach to transport were being questioned, notably by Peter Newman in a set of comparative studies of cities and their transport systems dating from the mid-1980s.

The British Government's White Paper on Transport marked a change in direction for transport planning in the UK. In the introduction to the White Paper, Prime Minister Tony Blair stated that:

> We recognise that we cannot simply build our way out of the problems we face. It would be environmentally irresponsible – and would not work.

A companion document to the White Paper called "Smarter Choices" researched the potential to scale up the small and scattered sustainable transport initiatives then occurring across Britain, and concluded that the comprehensive application of these techniques could reduce peak period car travel in urban areas by over 20%.

A similar study by the United States Federal Highway Administration, was also released in 2004 and also concluded that a more proactive approach to transportation demand was an important component of overall national transport strategy.

Environmental Impact

Transport systems are major emitters of greenhouse gases, responsible for 23% of world energy-related GHG emissions in 2004, with about three quarters coming from road vehicles. Currently 95% of transport energy comes from petroleum. Energy is consumed in the manufacture as well as the use of vehicles, and is embodied in transport infrastructure including roads, bridges and railways.

The first historical attempts of evaluating the life cycle environmental impact of vehicle is due to Theodore Von Karman. After decades in which all the analysis has been focused on emending the Von Karman model, Dewulf and Van Langenhove have introduced an model based on the second law of thermodynamics and exergy analysis. Chester and Orwath, have developed a similar model based on the first law that accounts the necessary costs for the infrastructure.

The Bus Rapid Transit of Metz uses a diesel-electric hybrid driving system, developed by Belgian Van Hool manufacturer.

The environmental impacts of transport can be reduced by reducing the weight of vehicles, sustainable styles of driving, reducing the friction of tires, encouraging electric and hybrid vehicles, improving the walking and cycling environment in cities, and by enhancing the role of public transport, especially electric rail.

Green vehicles are intended to have less environmental impact than equivalent standard vehicles, although when the environmental impact of a vehicle is assessed over the whole of its life cycle this may not be the case.

Electric vehicle technology (especially non-battery based vehicles, fuel cell vehicles) has the potential to reduce transport CO_2 emissions, depending on the embodied energy of the vehicle and the source of the electricity. The primary sources of electricity currently used in most countries (coal, gas, oil) mean that until world electricity production changes substantially, private electric cars will result in the same or higher production of CO2 than petrol equivalent vehicles. Battery-based electric vehicles may or may not be better in terms of GHG emissions then fossil-fuel based vehicles depending on several factors, such as battery type, capacity of the battery, life expectancy of the battery, etc.

The Online Electric Vehicle (OLEV), developed by the Korea Advanced Institute of Science and Technology (KAIST), is an electric vehicle that can be charged while stationary or driving, thus removing the need to stop at a charging station. The City of Gumi in South Korea runs a 24 km roundtrip along which the bus will receive 100 kW (136 horsepower) electricity at an 85% maximum power transmission efficiency rate while maintaining a 17 cm air gap between the underbody of the vehicle and the road surface. At that power, only a few sections of the road need embedded cables. Hybrid vehicles, which use an internal combustion engine combined with an electric engine to achieve better fuel efficiency than a regular combustion engine, are already common.

Natural gas is also used as a transport fuel but is a less promising, technology as it is still a fossil fuel and still has significant emissions (though lower then gasoline, diesel).

Brazil met 17% of its transport fuel needs from bioethanol in 2007, but the OECD has warned that the success of (first-generation) biofuels in Brazil is due to specific local circumstances.

Internationally, first-generation biofuels are forecast to have little or no impact on greenhouse emissions, at significantly higher cost than energy efficiency measures. The later generation biofuels however (2nd to 4th generation) do have significant environmental benefit, as they are no driving force for deforestation or struggle with the food vs fuel issue. Other renewable fuels include hydrogen, which (like drop-in biofuels) can be used in internal combustion vehicles, don't rely on any crops at all (instead it's produced using electricity) and even generates very little pollution when burned.

In practice there is a sliding scale of green transport depending on the sustainability of the option. Green vehicles are more fuel-efficient, but only in comparison with standard vehicles, and they still contribute to traffic congestion and road crashes. Well-patronised public transport networks based on traditional diesel buses use less fuel per passenger than private vehicles, and are generally safer and use less road space than private vehicles. Green public transport vehicles including electric trains, trams and electric buses combine the advantages of green vehicles with those of sustainable transport choices. Other transport choices with very low environmental impact are cycling and other human-powered vehicles, and animal powered transport. The most common green transport choice, with the least environmental impact is walking.

Transport and Social Sustainability

Cities with overbuilt roadways have experienced unintended consequences, linked to radical drops in public transport, walking, and cycling. In many cases, streets became void of "life." Stores, schools, government centers and libraries moved away from central cities, and residents who did not flee to the suburbs experienced a much reduced quality of public space and of public services. As schools were closed their mega-school replacements in outlying areas generated additional traffic; the number of cars on US roads between 7:15 and 8:15 a.m. increases 30% during the school year.

A tram in Melbourne, Australia.

Yet another impact was an increase in sedentary lifestyles, causing and complicating a national epidemic of obesity, and accompanying dramatically increased health care costs.

Cities

Cities are shaped by their transport systems. In The City in History, Lewis Mumford documented how the location and layout of cities was shaped around a walkable center, often located near a port or waterway, and with suburbs accessible by animal transport or, later, by rail or tram lines.

Futurama, an exhibit at the 1939 New York World's Fair.

In 1939, the New York World's Fair included a model of an imagined city, built around a car-based transport system. In this "greater and better world of tomorrow", residential, commercial and industrial areas were separated, and skyscrapers loomed over a network of urban motorways. These ideas captured the popular imagination, and are credited with influencing city planning from the 1940s to the 1970s.

Interstate 10 and Interstate 45 near downtown Houston, Texas.

The popularity of the car in the post-war era led to major changes in the structure and function of cities. There was some opposition to these changes at the time. The writings of Jane Jacobs, in particular The Death and Life of Great American Cities provide a poignant reminder of what was lost in this transformation, and a record of community efforts to resist these changes. Lewis Mumford asked "is the city for cars or for people?" Donald Appleyard documented the consequences for communities of increasing car traffic in "The View from the Road" and in the UK, Mayer Hillman first published research into the impacts of traffic on child independent mobility in 1971. Despite these notes of caution, trends in car ownership, car use and fuel consumption continued steeply upward throughout the post-war period.

Mainstream transport planning in Europe has, by contrast, never been based on assumptions that the private car was the best or only solution for urban mobility. For example, the Dutch Transport Structure Scheme has since the 1970s required that demand for additional vehicle capacity only be met "if the contribution to societal welfare is positive", and since 1990 has included an explicit target to halve the rate of growth in vehicle traffic. Some cities outside Europe have also consistently linked transport to sustainability and to land-use planning, notably Curitiba, Brazil, Portland, Oregon and Vancouver, Canada.

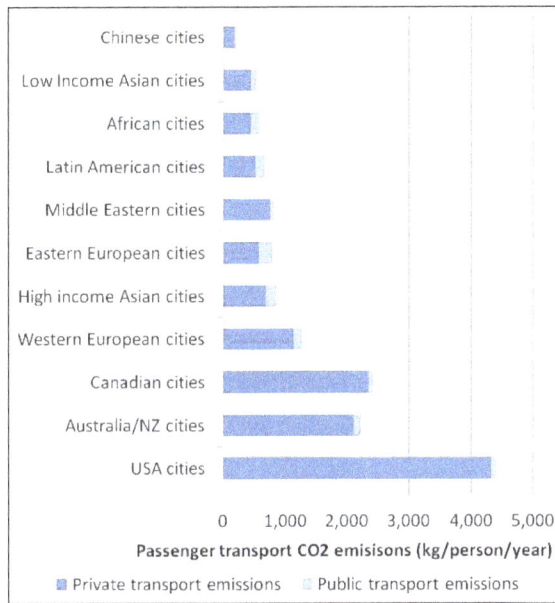

Greenhouse gas emissions from transport vary widely, even for cities of comparable wealth.

There are major differences in transport energy consumption between cities; an average U.S. urban dweller uses 24 times more energy annually for private transport than a Chinese urban resident, and almost four times as much as a European urban dweller. These differences cannot be explained by wealth alone but are closely linked to the rates of walking, cycling, and public transport use and to enduring features of the city including urban density and urban design.

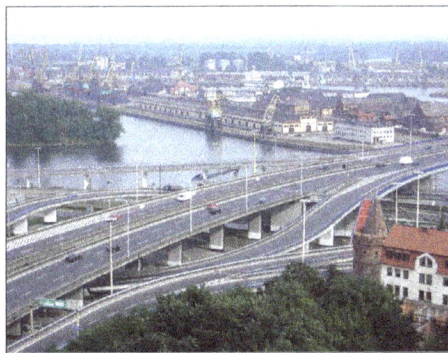

A bypass the Old Town in Szczecin, Poland.

The cities and nations that have invested most heavily in car-based transport systems are now the least environmentally sustainable, as measured by per capita fossil fuel use. The social and economic sustainability of car-based transportation engineering has also been questioned. Within the United States, residents of sprawling cities make more frequent and longer car trips, while residents of traditional urban neighbourhoods make a similar number of trips, but travel shorter distances and walk, cycle and use transit more often. It has been calculated that New York residents save $19 billion each year simply by owning fewer cars and driving less than the average American. A less car intensive means of urban transport is carsharing, which is becoming popular in North America and Europe, and according to The Economist, carsharing can reduce car ownership at an estimated rate of one rental car replacing 15 owned vehicles. Car sharing has also begun in the developing world, where traffic and urban density is often worse than in developed countries.

Companies like Zoom in India, eHi in China, and Carrot in Mexico, are bringing car-sharing to developing countries in an effort to reduce car-related pollution, ameliorate traffic, and expand the number of people who have access to cars.

The European Commission adopted the Action Plan on urban mobility on 2009-09-30 for sustainable urban mobility. The European Commission will conduct a review of the implementation of the Action Plan in the year 2012, and will assess the need for further action. In 2007, 72% of the European population lived in urban areas, which are key to growth and employment. Cities need efficient transport systems to support their economy and the welfare of their inhabitants. Around 85% of the EU's GDP is generated in cities. Urban areas face today the challenge of making transport sustainable in environmental (CO_2, air pollution, noise) and competitiveness (congestion) terms while at the same time addressing social concerns. These range from the need to respond to health problems and demographic trends, fostering economic and social cohesion to taking into account the needs of persons with reduced mobility, families and children.

Policies and Governance

Seven sustainable transportations (Prague).

Sustainable transport policies have their greatest impact at the city level. Outside Western Europe, cities which have consistently included sustainability as a key consideration in transport and land use planning include Curitiba, Brazil; Bogota, Colombia; Portland, Oregon; and Vancouver, Canada. The state of Victoria, Australia passed legislation in 2010 – the Transport Integration Act – to compel its transport agencies to actively consider sustainability issues including climate change impacts in transport policy, planning and operations. Many other cities throughout the world have recognised the need to link sustainability and transport policies, for example by joining the Cities for Climate Protection program.

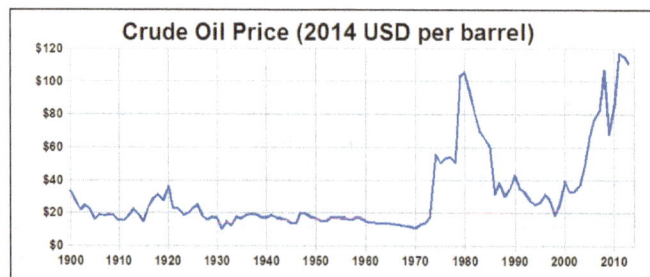

Oil price trend, 1939–2007, both nominal and adjusted to inflation.

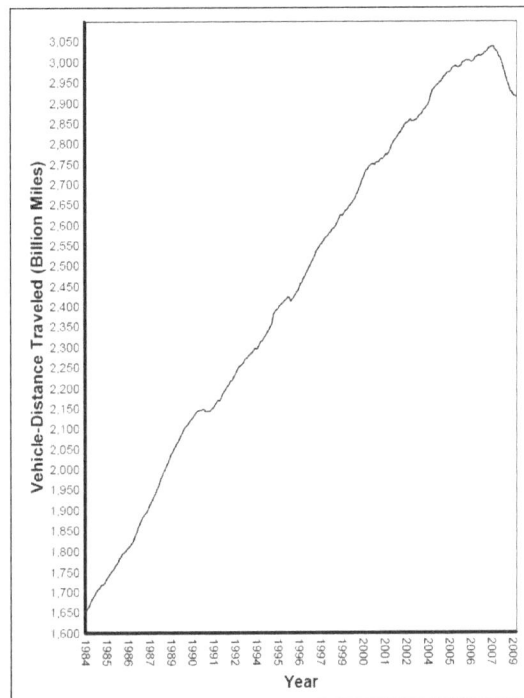

Vehicle-miles traveled in the United States up to March 2009.

Community and Grassroots Action

Sustainable transport is fundamentally a grassroots movement, albeit one which is now recognised as of citywide, national and international significance.

Whereas it started as a movement driven by environmental concerns, over these last years there has been increased emphasis on social equity and fairness issues, and in particular the need to ensure proper access and services for lower income groups and people with mobility limitations, including the fast-growing population of older citizens. Many of the people exposed to the most vehicle noise, pollution and safety risk have been those who do not own, or cannot drive cars, and those for whom the cost of car ownership causes a severe financial burden.

An organization called Greenxc started in 2011 created a national awareness campaign in the United States encouraging people to carpool by ride-sharing cross country stopping over at various destinations along the way and documenting their travel through video footage, posts and photography. Ride-sharing reduces individual's carbon footprint by allowing several people to use one car instead of everyone using individual cars.

Tools and Incentives

Several European countries are opening up financial incentives that support more sustainable modes of transport. The European Cyclists' Federation, which focuses on daily cycling for transport, has created a document containing a non-complete overview. In the UK, employers have for many years been providing employees with financial incentives. The employee leases or borrows a bike that the employer has purchased. You can also get other support. The scheme is beneficial for the employee who saves money and gets an incentive to get exercise integrated in the daily routine.

The employer can expect a tax deduction, lower sick leave and less pressure on parking spaces for cars. Since 2010, there has been a scheme in Iceland (Samgöngugreiðslur) where those who do not drive a car to work, get paid a lump of money monthly. An employee must sign a statement not to use a car for work more often than one day a week, or 20% of the days for a period. Some employers pay fixed amounts based on trust. Other employers reimburse the expenses for repairs on bicycles, period-tickets for public transport and the like. Since 2013, amounts up to ISK 8000 per month have been tax-free. Most major workplaces offer this, and a significant proportion of employees use the scheme. From the year 2019 half the amount is tax-free if the employee sings a contract not to use a car to work for more than 40% of the days of the contract period.

Greenwashing

The term *green transport* is often used as a greenwash marketing technique for products which are not proven to make a positive contribution to environmental sustainability. Such claims can be legally challenged. For instance Norway's consumer ombudsman has targeted automakers who claim that their cars are "green", "clean" or "environmentally friendly". Manufacturers risk fines if they fail to drop the words. The Australian Competition and Consumer Commission (ACCC) describes *green* claims on products as *very vague, inviting consumers to give a wide range of meanings to the claim, which risks misleading them.* In 2008 the ACCC forced a car retailer to stop its *green* marketing of Saab cars, which was found by the Australian Federal Court as *misleading.*

SUSTAINABLE LIVING

Sustainable living describes a lifestyle that attempts to reduce an individual's or society's use of the Earth's natural resources, and one's personal resources. It is often called as "earth harmony living" or "net zero living". Its practitioners often attempt to reduce their carbon footprint by altering their methods of transportation, energy consumption, and diet. Its proponents aim to conduct their lives in ways that are consistent with sustainability, naturally balanced, and respectful of humanity's symbiotic relationship with the Earth's natural ecology. The practice and general philosophy of ecological living closely follows the overall principles of sustainable development.

Lester R. Brown, a prominent environmentalist and founder of the Worldwatch Institute and Earth Policy Institute, describes sustainable living in the twenty-first century as "shifting to a renewable energy-based, reuse/recycle economy with a diversified transport system." Derrick Jensen ("the poet-philosopher of the ecological movement"), a celebrated American author, radical environmentalist and prominent critic of mainstream environmentalism argues that "industrial civilization is not and can never be sustainable". From this statement, the natural conclusion is that sustainable living is at odds with industrialization. Thus, practitioners of the philosophy potentially face the challenge of living in an industrial society and adapting alternative norms, technologies, or practices.

Additionally, practical ecovillage builders like Living Villages maintain that the shift to alternative technologies will only be successful if the resultant built environment is attractive to a local culture and can be maintained and adapted as necessary over multiple generations.

Sustainable living is fundamentally the application of sustainability to lifestyle choice and decisions. One conception of sustainable living expresses what it means in triple-bottom-line terms as meeting present ecological, societal, and economical needs without compromising these factors for future generations. Another broader conception describes sustainable living in terms of four interconnected *social* domains: economics, ecology, politics and culture. In the first conception, sustainable living can be described as living within the innate carrying capacities defined by these factors. In the second or Circles of Sustainability conception, sustainable living can be described as negotiating the relationships of needs within limits across all the interconnected domains of social life, including consequences for future human generations and non-human species.

Sustainable design and sustainable development are critical factors to sustainable living. Sustainable design encompasses the development of appropriate technology, which is a staple of sustainable living practices. Sustainable development in turn is the use of these technologies in infrastructure. Sustainable architecture and agriculture are the most common examples of this practice.

Shelter

On a global scale, shelter is associated with about 25% of the greenhouse gas emissions embodied in household purchases and 26% of households' land use.

An example of ecological housing.

Sustainable homes are built using sustainable methods, materials, and facilitate green practices, enabling a more sustainable lifestyle. Their construction and maintenance have neutral impacts on the Earth. Often, if necessary, they are close in proximity to essential services such as grocery stores, schools, daycares, work, or public transit making it possible to commit to sustainable transportation choices. Sometimes, they are off-the-grid homes that do not require any public energy, water, or sewer service.

If not off-the-grid, sustainable homes may be linked to a grid supplied by a power plant that is using sustainable power sources, buying power as is normal convention. Additionally, sustainable homes may be connected to a grid, but generate their own electricity through renewable means and sell any excess to a utility. There are two common methods to approaching this option: net metering and double metering.

Net metering uses the common meter that is installed in most homes, running forward when power is used from the grid, and running backward when power is put into the grid (which allows them to "net" out their total energy use, putting excess energy into the grid when not needed, and using energy from the grid during peak hours, when you may not be able to produce enough immediately). Power companies can quickly purchase the power that is put back into the grid, as it is being produced. Double metering involves installing two meters: one measuring electricity consumed, the other measuring electricity created. Additionally, or in place of selling their renewable energy, sustainable home owners may choose to bank their excess energy by using it to charge batteries. This gives them the option to use the power later during less favorable power-generating times (i.e. night-time, when there has been no wind, etc.), and to be completely independent of the electrical grid.

Sustainably designed houses are generally sited so as to create as little of a negative impact on the surrounding ecosystem as possible, oriented to the sun so that it creates the best possible micro-climate (typically, the long axis of the house or building should be oriented east-west), and provide natural shading or wind barriers where and when needed, among many other considerations. The design of a sustainable shelter affords the options it has later (i.e. using passive solar lighting and heating, creating temperature buffer zones by adding porches, deep overhangs to help create favorable microclimates, etc.). Sustainably constructed houses involve environmentally friendly management of waste building materials such as recycling and composting, use non-toxic and renewable, recycled, reclaimed, or low-impact production materials that have been created and treated in a sustainable fashion (such as using organic or water-based finishes), use as much locally available materials and tools as possible so as to reduce the need for transportation, and use low-impact production methods (methods that minimize effects on the environment).

Many materials can be considered a "green" material until its background is revealed. Any material that has used toxic or carcinogenic chemicals in its treatment or manufacturing (such as formaldehyde in glues used in woodworking), has traveled extensively from its source or manufacturer, or has been cultivated or harvested in an unsustainable manner might not be considered green. In order for any material to be considered green, it must be resource efficient, not compromise indoor air quality or water conservation, and be energy efficient (both in processing and when in use in the shelter). Resource efficiency can be achieved by using as much recycled content, reusable or recyclable content, materials that employ recycled or recyclable packaging, locally available material, salvaged or remanufactured material, material that employs resource efficient manufacturing, and long-lasting material as possible.

Sustainable Building Materials

Some building materials might be considered "sustainable" by some definitions and under some conditions. For example, wood might be thought of as sustainable if it is grown using sustainable forest management, processed using sustainable energy. delivered by sustainable transport, etc. Under different conditions, however, it might not be considered as sustainable. The following materials might be considered as sustainable under certain conditions, based on a life-cycle assessment.

- Adobe
- Bamboo
- Cellulose insulation

- Cob
- Composite wood (when made from reclaimed hardwood sawdust and reclaimed or recycled plastic)

- Compressed earth block
- Cordwood
- Cork
- Hemp
- Insulating concrete forms
- Lime render
- Linoleum
- Lumber from Forest Stewardship Council approved sources
- Natural Rubber
- Natural fiber (coir, wool, jute, etc.)
- Organic cotton insulation

- Papercrete
- Rammed earth
- Reclaimed stone
- Reclaimed brick
- Recycled metal
- Recycled concrete
- Recycled paper
- Soy-based adhesive
- Soy insulation
- Straw Bale
- Structural insulated panel
- Wood

Insulation of a sustainable home is important because of the energy it conserves throughout the life of the home. Well insulated walls and lofts using green materials are a must as it reduces or in combination with a house that is well designed, eliminates the need for heating and cooling altogether. Installation of insulation varies according to the type of insulation being used. Typically, lofts are insulated by strips of insulating material laid between rafters. Walls with cavities are done in much the same manner. For walls that do not have cavities behind them, solid-wall insulation may be necessary which can decrease internal space and can be expensive to install. Energy-efficient windows are another important factor in insulation. Simply assuring that windows (and doors) are well sealed greatly reduces energy loss in a home. Double or Triple glazed windows are the typical method to insulating windows, trapping gas or creating a vacuum between two or three panes of glass allowing heat to be trapped inside or out. Low-emissivity or Low-E glass is another option for window insulation. It is a coating on windowpanes of a thin, transparent layer of metal oxide and works by reflecting heat back to its source, keeping the interior warm during the winter and cool during the summer. Simply hanging heavy-backed curtains in front of windows may also help their insulation. "Superwindows," in Natural Capitalism: Creating the Next Industrial Revolution, became available in the 1980s and use a combination of many available technologies, including two to three transparent low-e coatings, multiple panes of glass, and a heavy gas filling. Although more expensive, they are said to be able to insulate four and a half times better than a typical double-glazed windows.

Equipping roofs with highly reflective material (such as aluminum) increases a roof's albedo and will help reduce the amount of heat it absorbs, hence, the amount of energy needed to cool the building it is on. Green roofs or "living roofs" are a popular choice for thermally insulating a building. They are also popular for their ability to catch storm-water runoff and when in the broader picture of a community, reduce the heat island effect thereby reducing energy costs of the entire area. It is arguable that they are able to replace the physical "footprint" that the building creates, helping reduce the adverse environmental impacts of the building's presence.

Energy efficiency and water conservation are also major considerations in sustainable housing. If using appliances, computers, HVAC systems, electronics, or lighting the sustainable-minded often look for an Energy Star label, which is government-backed and holds stricter regulations in energy and water efficiency than is required by law. Ideally, a sustainable shelter should be able to completely run the appliances it uses using renewable energy and should strive to have a neutral impact on the Earth's water sources.

Greywater, including water from washing machines, sinks, showers, and baths may be reused in landscape irrigation and toilets as a method of water conservation. Likewise, rainwater harvesting from storm-water runoff is also a sustainable method to conserve water use in a sustainable shelter. Sustainable Urban Drainage Systems replicate the natural systems that clean water in wildlife and implement them in a city's drainage system so as to minimize contaminated water and unnatural rates of runoff into the environment.

Power

As mentioned under Shelter, some sustainable households may choose to produce their own renewable energy, while others may choose to purchase it through the grid from a power company that harnesses sustainable sources (are the methods of metering the production and consumption of electricity in a household). Purchasing sustainable energy, however, may simply not be possible in some locations due to its limited availability. 6 out of the 50 states in the US do not offer green energy, for example. For those that do, its consumers typically buy a fixed amount or a percentage of their monthly consumption from a company of their choice and the bought green energy is fed into the entire national grid. Technically, in this case, the green energy is not being fed directly to the household that buys it. In this case, it is possible that the amount of green electricity that the buying household receives is a small fraction of their total incoming electricity. This may or may not depend on the amount being purchased. The purpose of buying green electricity is to support their utility's effort in producing sustainable energy. Producing sustainable energy on an individual household or community basis is much more flexible, but can still be limited in the richness of the sources that the location may afford (some locations may not be rich in renewable energy sources while others may have an abundance of it).

Sustainable urban design and innovation: Photovoltaic ombrière SUDI is an autonomous and mobile station that replenishes energy for electric vehicles using solar energy.

When generating renewable energy and feeding it back into the grid (in participating countries such as the US and Germany), producing households are typically paid at least the full standard electricity rate by their utility and are also given separate renewable energy credits that they can then sell to their utility, additionally (utilities are interested in buying these renewable energy credits because it allows them to claim that they produce renewable energy). In some special cases, producing households may be paid up to four times the standard electricity rate, but this is not common.

An installation of solar panels in rural Mongolia.

Solar power harnesses the energy of the sun to make electricity. Two typical methods for converting solar energy into electricity are photo-voltaic cells that are organized into panels and concentrated solar power, which uses mirrors to concentrate sunlight to either heat a fluid that runs an electrical generator via a steam turbine or heat engine, or to simply cast onto photo-voltaic cells. The energy created by photo-voltaic cells is a direct current and has to be converted to alternating current before it can be used in a household. At this point, users can choose to either store this direct current in batteries for later use, or use an AC/DC inverter for immediate use. To get the best out of a solar panel, the angle of incidence of the sun should be between 20 and 50 degrees. Solar power via photo-voltaic cells are usually the most expensive method to harnessing renewable energy, but is falling in price as technology advances and public interest increases. It has the advantages of being portable, easy to use on an individual basis, readily available for government grants and incentives, and being flexible regarding location (though it is most efficient when used in hot, arid areas since they tend to be the most sunny). For those that are lucky, affordable rental schemes may be found. Concentrated solar power plants are typically used on more of a community scale rather than an individual household scale, because of the amount of energy they are able to harness but can be done on an individual scale with a parabolic reflector.

Solar thermal energy is harnessed by collecting direct heat from the sun. One of the most common ways that this method is used by households is through solar water heating. In a broad perspective, these systems involve well insulated tanks for storage and collectors, are either passive or active systems (active systems have pumps that continuously circulate water through the collectors and storage tank) and in active systems, involve either directly heating the water that will be used or heating a non-freezing heat-transfer fluid that then heats the water that will be used. Passive systems are cheaper than active systems since they do not require a pumping system (instead, they take advantage of the natural movement of hot water rising above cold water to cycle the water being used through the collector and storage tank).

Other methods of harnessing solar power are solar space heating (for heating internal building spaces), solar drying (for drying wood chips, fruits, grains, etc.), solar cookers, solar distillers, and other passive solar technologies (simply, harnessing sunlight without any mechanical means).

Wind power is harnessed through turbines, set on tall towers (typically 20' or 6 m with 10' or 3 m diameter blades for an individual household's needs) that power a generator that creates electricity. They typically require an average of wind speed of 9 mi/hr (14 km/hr) to be worth their investment and are capable of paying for themselves within their lifetimes. Wind turbines in urban areas usually need to be mounted at least 30' (10 m) in the air to receive enough wind and to be void of nearby obstructions (such as neighboring buildings). Mounting a wind turbine may also require permission from authorities. Wind turbines have been criticized for the noise they produce, their appearance, and the argument that they can affect the migratory patterns of birds (their blades obstruct passage in the sky). Wind turbines are much more feasible for those living in rural areas and are one of the most cost-effective forms of renewable energy per kilowatt, approaching the cost of fossil fuels, and have quick paybacks.

For those that have a body of water flowing at an adequate speed (or falling from an adequate height) on their property, hydroelectricity may be an option. On a large scale, hydroelectricity, in the form of dams, has adverse environmental and social impacts. When on a small scale, however, in the form of single turbines, hydroelectricity is very sustainable. Single water turbines or even a group of single turbines are not environmentally or socially disruptive. On an individual household basis, single turbines are the probably the only economically feasible route (but can have high paybacks and is one of the most efficient methods of renewable energy production). It is more common for an eco-village to use this method rather than a singular household.

Geothermal energy production involves harnessing the hot water or steam below the earth's surface, in reservoirs, to produce energy. Because the hot water or steam that is used is reinjected back into the reservoir, this source is considered sustainable. However, those that plan on getting their electricity from this source should be aware that there is controversy over the lifespan of each geothermal reservoir as some believe that their lifespans are naturally limited (they cool down over time, making geothermal energy production there eventually impossible). This method is often large scale as the system required to harness geothermal energy can be complex and requires deep drilling equipment. There do exist small individual scale geothermal operations, however, which harness reservoirs very close to the Earth's surface, avoiding the need for extensive drilling and sometimes even taking advantage of lakes or ponds where there is already a depression. In this case, the heat is captured and sent to a geothermal heat pump system located inside the shelter or facility that needs it (often, this heat is used directly to warm a greenhouse during the colder months). Although geothermal energy is available everywhere on Earth, practicality and cost-effectiveness varies, directly related to the depth required to reach reservoirs. Places such as the Philippines, Hawaii, Alaska, Iceland, California, and Nevada have geothermal reservoirs closer to the Earth's surface, making its production cost-effective.

Biomass power is created when any biological matter is burned as fuel. As with the case of using green materials in a household, it is best to use as much locally available material as possible so as to reduce the carbon footprint created by transportation. Although burning biomass for fuel releases carbon dioxide, sulfur compounds, and nitrogen compounds into the atmosphere, a major concern in a sustainable lifestyle, the amount that is released is sustainable (it will not contribute to a rise in carbon dioxide levels in the atmosphere). This is because the biological matter that is being burned releases the same amount of carbon dioxide that it consumed during its lifetime. However, burning biodiesel and bioethanol when created from virgin material, is increasingly controversial and may or may not be considered sustainable because it inadvertently increases global poverty, the clearing of more land for new agriculture fields (the source of the biofuel is also the same source of food), and may use unsustainable growing methods (such as the use of environmentally harmful pesticides and fertilizers).

List of organic matter than can be burned for fuel:

- Bagasse
- Biogas
- Manure
- Straw
- Used vegetable oil
- Wood

Digestion of organic material to produce methane is becoming an increasingly popular method of biomass energy production. Materials such as waste sludge can be digested to release methane gas that can then be burnt to produce electricity. Methane gas is also a natural by-product of landfills, full of decomposing waste, and can be harnessed here to produce electricity as well. The advantage in burning methane gas is that is prevents the methane from being released into the atmosphere, exacerbating the greenhouse effect. Although this method of biomass energy production is typically large scale (done in landfills), it can be done on a smaller individual or community scale as well.

Food

Globally, food accounts for 48% and 70% of household environmental impacts on land and water resources respectively, with consumption of meat, dairy and processed food rising quickly with income.

Environmental Impacts of Industrial Agriculture

Industrial agricultural production is highly resource and energy intensive. Industrial agriculture systems typically require heavy irrigation, extensive pesticide and fertilizer application, intensive tillage, concentrated monoculture production, and other continual inputs. As a result of these industrial farming conditions, today's mounting environmental stresses are further exacerbated. These stresses include: declining water tables, chemical leaching, chemical runoff, soil erosion, land degradation, loss in biodiversity, and other ecological concerns.

Conventional Food Distribution and Long Distance Transport

Conventional food distribution and long distance transport are additionally resource and energy exhaustive. Substantial climate-disrupting carbon emissions, boosted by the transport of food over long distances, are of growing concern as the world faces such global crisis as natural resource depletion, peak oil and climate change. "The average American meal currently costs about 1500 miles, and takes about 10 calories of oil and other fossil fuels to produce a single calorie of food."

Local and Seasonal Foods

A more sustainable means of acquiring food is to purchase locally and seasonally. Buying food from local farmers reduces carbon output, caused by long-distance food transport, and stimulates the local economy. Local, small-scale farming operations also typically utilize more sustainable methods of agriculture than conventional industrial farming systems such as decreased tillage, nutrient cycling, fostered biodiversity and reduced chemical pesticide and fertilizer applications. Adapting a more regional, seasonally based diet is more sustainable as it entails purchasing less energy and resource demanding produce that naturally grow within a local area and require no long-distance transport. These vegetables and fruits are also grown and harvested within their suitable growing season. Thus, seasonal food farming does not require energy intensive greenhouse production, extensive irrigation, plastic packaging and long-distance transport from importing non-regional foods, and other environmental stressors. Local, seasonal produce is typically fresher, unprocessed and argued to be more nutritious. Local produce also contains less to no chemical residues from applications required for long-distance shipping and handling. Farmers' markets, public events where local small-scale farmers gather and sell their produce, are a good source for obtaining local food and knowledge about local farming productions. As well as promoting localization of food, farmers markets are a central gathering place for community interaction. Another way to become involved in regional food distribution is by joining a local community-supported agriculture (CSA). A CSA consists of a community of growers and consumers who pledge to support a farming operation while equally sharing the risks and benefits of food production. CSA's usually involve a system of weekly pick-ups of locally farmed vegetables and fruits, sometimes including dairy products, meat and special food items such as baked goods. Considering the rising environmental crisis, the United States and much of the world is facing immense vulnerability to famine. Local food production ensures food security if potential transportation disruptions and climatic, economical, and sociopolitical disasters were to occur.

Reducing Meat Consumption

Industrial meat production also involves high environmental costs such as land degradation, soil erosion and depletion of natural resources, especially pertaining to water and food. Mass meat production increase the amount of methane in the atmosphere. Reducing meat consumption, perhaps to a few meals a week, or adopting a vegetarian or vegan diet, alleviates the demand for environmentally damaging industrial meat production. Buying and consuming organically raised, free range or grass fed meat is another alternative towards more sustainable meat consumption.

Organic Farming

Purchasing and supporting organic products is another fundamental contribution to sustainable living. Organic farming is a rapidly emerging trend in the food industry and in the web of sustainability. According to the USDA National Organic Standards Board (NOSB), organic agriculture is defined as "an ecological production management system that promotes and enhances biodiversity, biological cycles, and soil biological activity. It is based on minimal use of off-farm inputs and on management practices that restore, maintain, or enhance ecological harmony. The primary goal of organic agriculture is to optimize the health and productivity of interdependent communities of soil life, plants, animals and people." Upon sustaining these goals, organic agriculture uses techniques such as crop rotation, permaculture, compost, green manure and biological pest control. In addition, organic farming prohibits or strictly limits the use of manufactured fertilizers and pesticides, plant growth regulators such as hormones, livestock antibiotics, food additives and genetically modified organisms. Organically farmed products include vegetables, fruit, grains, herbs, meat, dairy, eggs, fibers, and flowers.

Urban Gardening

Edible landscaping: a vegetable garden incorporated by the local residents into a roadside park.

In addition to local, small-scale farms, there has been a recent emergence in urban agriculture expanding from community gardens to private home gardens. With this trend, both farmers and ordinary people are becoming involved in food production. A network of urban farming systems helps to further ensure regional food security and encourages self-sufficiency and cooperative interdependence within communities. With every bite of food raised from urban gardens, negative environmental impacts are reduced in numerous ways. For instance, vegetables and fruits raised within small-scale gardens and farms are not grown with tremendous applications of nitrogen fertilizer required for industrial agricultural operations. The nitrogen fertilizers cause toxic chemical leaching and runoff that enters our water tables. Nitrogen fertilizer also produces nitrous oxide, a more damaging greenhouse gas than carbon dioxide. Local, community-grown food also requires no imported, long-distance transport which further depletes our fossil fuel reserves. In developing more efficiency per land acre, urban gardens can be started in a wide variety of areas: in vacant lots, public parks, private yards, church and school yards, on roof tops (roof-top gardens), and many other places. Communities can work together in changing zoning limitations in order for public and private gardens to be permissible. Aesthetically pleasing edible landscaping plants can also be incorporated into city landscaping such as blueberry bushes, grapevines trained on an arbor, pecan trees, etc. With as small a scale as home or community farming, sustainable and organic farming methods can easily be utilized. Such sustainable, organic farming techniques include: composting, biological pest control, crop rotation, mulching, drip irrigation, nutrient cycling and permaculture.

Food Preservation and Storage

Preserving and storing foods reduces reliance on long-distance transported food and the market industry. Home-grown foods can be preserved and stored outside of their growing season and continually consumed throughout the year, enhancing self-sufficiency and independence from the supermarket. Food can be preserved and saved by dehydration, freezing, vacuum packing, canning, bottling, pickling and jellying.

Transportation

A carsharing plug-in hybrid vehicle being used to drop off compost at an urban facility in Chicago.

With rising peak oil concerns, climate warming exacerbated by carbon emissions and high energy prices, the conventional automobile industry is becoming less and less feasible to the conversation of sustainability. Revisions of urban transport systems that foster mobility, low-cost transportation and healthier urban environments are needed. Such urban transport systems should consist of a combination of rail transport, bus transport, bicycle pathways and pedestrian walkways. Public transport systems such as underground rail systems and bus transit systems shift huge numbers of people away from reliance on car mobilization and dramatically reduce the rate of carbon emissions caused by automobile transport. Carpooling is another alternative for reducing oil consumption and carbon emissions by transit.

In comparison to automobiles, bicycles are a paragon of energy efficient personal transportation with the bicycle roughly 50 times more energy efficient than driving. Bicycles increase mobility while alleviating congestion, lowering air and noise pollution, and increasing physical exercise. Most importantly, they do not emit climate-disturbing carbon dioxide. Bike-sharing programs are beginning to boom throughout the world and are modeled in leading cities such as Paris, Amsterdam and London. Bike-sharing programs offer kiosks and docking stations that supply hundreds to thousands of bikes for rental throughout a city through small deposits or affordable memberships.

A recent boom has occurred in electric bikes especially in China and other Asian countries. Electric bikes are similar to plug-in hybrid vehicles in that they are battery powered and can be plugged into the provincial electric grid for recharging as needed. In contrast to plug-in hybrid cars, electric bikes do not directly use any fossil fuels. Adequate sustainable urban transportation is dependent upon proper city infrastructure and planning that incorporates efficient public transit along with bicycle and pedestrian-friendly pathways.

Water

A major factor of sustainable living involves that which no human can live without, water. Unsustainable water use has far reaching implications for humankind. Currently, humans use one-fourth of the Earth's total fresh water in natural circulation, and over half the accessible runoff. Additionally, population growth and water demand is ever increasing. Thus, it is necessary to use available water more efficiently. In sustainable living, one can use water more sustainably through a series of simple, everyday measures. These measures involve considering indoor home appliance efficiency, outdoor water use, and daily water use awareness.

Indoor Home Appliances

Housing and commercial buildings account for 12 percent of America's freshwater withdrawals. A typical American single family home uses about 70 US gallons (260 L) per person per day indoors. This use can be reduced by simple alterations in behavior and upgrades to appliance quality.

Toilets

Toilets accounted for almost 30% of residential indoor water use in the United States in 1999. One flush of a standard U.S. toilet requires more water than most individuals, and many families, in the world use for all their needs in an entire day. A home's toilet water sustainability can be improved in one of two ways: improving the current toilet or installing a more efficient toilet. To improve the current toilet, one possible method is to put weighted plastic bottles in the toilet tank. Also, there are inexpensive tank banks or float booster available for purchase. A tank bank is a plastic bag to be filled with water and hung in the toilet tank. A float booster attaches underneath the float ball of pre-1986 three and a half gallon capacity toilets. It allows these toilets to operate at the same valve and float setting but significantly reduces their water level, saving between one and one and a third gallons of water per flush. A major waste of water in existing toilets is leaks. A slow toilet leak is undetectable to the eye, but can waste hundreds of gallons each month. One way to check this is to put food dye in the tank, and to see if the water in the toilet bowl turns the same color. In the event of a leaky flapper, one can replace it with an adjustable toilet flapper, which allows self-adjustment of the amount of water per flush.

In installing a new toilet there are a number of options to obtain the most water efficient model. A low flush toilet uses one to two gallons per flush. Traditionally, toilets use three to five gallons per flush. If an eighteen-liter per flush toilet is removed and a six-liter per flush toilet is put in its place, 70% of the water flushed will be saved while the overall indoor water use by will be reduced by 30%. It is possible to have a toilet that uses no water. A composting toilet treats human waste through composting and dehydration, producing a valuable soil additive. These toilets feature a two-compartment bowl to separate urine from feces. The urine can be collected or sold as fertilizer. The feces can be dried and bagged or composted. These toilets cost scarcely more than regularly installed toilets and do not require a sewer hookup. In addition to providing valuable fertilizer, these toilets are highly sustainable because they save sewage collection and treatment, as well as lessen agricultural costs and improve topsoil.

Additionally, one can reduce toilet water sustainability by limiting total toilet flushing. For instance, instead of flushing small wastes, such as tissues, one can dispose of these items in the trash or compost.

Showers

On average, showers were 18% of U.S. indoor water use in 1999, at 6–8 US gallons (23–30 L) per minute traditionally in America. A simple method to reduce this use is to switch to low-flow, high-performance showerheads. These showerheads use only 1.0-1.5 gpm or less. An alternative to replacing the showerhead is to install a converter. This device arrests a running shower upon reaching the desired temperature. Solar water heaters can be used to obtain optimal water temperature, and are more sustainable because they reduce dependence on fossil fuels. To lessen excess water use, water pipes can be insulated with pre-slit foam pipe insulation. This insulation decreases hot water generation time. A simple, straightforward method to conserve water when showering is to take shorter showers. One method to accomplish this is to turn off the water when it is not necessary (such as while lathering) and resuming the shower when water is necessary. This can be facilitated when the plumbing or showerhead allow turning off the water without disrupting the desired temperature setting (common in the UK but not the United States).

Dishwashers and Sinks

On average, sinks were 15% of U.S. indoor water use in 1999. There are, however, easy methods to rectify excessive water loss. Available for purchase is a screw-on aerator. This device works by combining water with air thus generating a frothy substance with greater perceived volume, reducing water use by half. Additionally, there is a flip-valve available that allows flow to be turned off and back on at the previously reached temperature. Finally, a laminar flow device creates a 1.5-2.4 gpm stream of water that reduces water use by half, but can be turned to normal water level when optimal.

In addition to buying the above devices, one can live more sustainably by checking sinks for leaks, and fixing these links if they exist. According to the EPA, "A small drip from a worn faucet washer can waste 20 gallons of water per day, while larger leaks can waste hundreds of gallons". When washing dishes by hand, it is not necessary to leave the water running for rinsing, and it is more efficient to rinse dishes simultaneously.

On average, dishwashing consumes 1% of indoor water use. When using a dishwasher, water can be conserved by only running the machine when it is full. Some have a "low flow" setting to use less water per wash cycle. Enzymatic detergents clean dishes more efficiently and more successfully with a smaller amount of water at a lower temperature.

Washing Machines

On average, 23% of U.S. indoor water use in 1999 was due to clothes washing. In contrast to other machines, American washing machines have changed little to become more sustainable. A typical washing machine has a vertical-axis design, in which clothes are agitated in a tubful of water. Horizontal-axis machines, in contrast, put less water into the bottom of the rub and rotate clothes through it. These machines are more efficient in terms of soap use and clothing stability.

Outdoor Water Use

There are a number of ways one can incorporate a personal yard, roof, and garden in more sustainable living. While conserving water is a major element of sustainability, so is sequestering water.

Conserving Water

In planning a yard and garden space, it is most sustainable to consider the plants, soil, and available water. Drought resistant shrubs, plants, and grasses require a smaller amount of water in comparison to more traditional species. Additionally, native plants (as opposed to herbaceous perennials) will use a smaller supply of water and have a heightened resistance to plant diseases of the area. Xeriscaping is a technique that selects drought-tolerant plants and accounts for endemic features such as slope, soil type, and native plant range. It can reduce landscape water use by 50 – 70%, while providing habitat space for wildlife. Plants on slopes help reduce runoff by slowing and absorbing accumulated rainfall. Grouping plants by watering needs further reduces water waste.

After planting, placing a circumference of mulch surrounding plants functions to lessen evaporation. To do this, firmly press two to four inches of organic matter along the plant's dripline. This prevents water runoff. When watering, consider the range of sprinklers; watering paved areas is unnecessary. Additionally, to conserve the maximum amount of water, watering should be carried out during early mornings on non-windy days to reduce water loss to evaporation. Drip-irrigation systems and soaker hoses are a more sustainable alternative to the traditional sprinkler system. Drip-irrigation systems employ small gaps at standard distances in a hose, leading to the slow trickle of water droplets which percolate the soil over a protracted period. These systems use 30 – 50% less water than conventional methods. Soaker hoses help to reduce water use by up to 90%. They connect to a garden hose and lay along the row of plants under a layer of mulch. A layer of organic material added to the soil helps to increase its absorption and water retention; previously planted areas can be covered with compost.

In caring for a lawn, there are a number of measures that can increase the sustainability of lawn maintenance techniques. A primary aspect of lawn care is watering. To conserve water, it is important to only water when necessary, and to deep soak when watering. Additionally, a lawn may be left to go dormant, renewing after a dry spell to its original vitality.

Sequestering Water

A common method of water sequestrations is rainwater harvesting, which incorporates the collection and storage of rain. Primarily, the rain is obtained from a roof, and stored on the ground in catchment tanks. Water sequestration varies based on extent, cost, and complexity. A simple method involves a single barrel at the bottom of a downspout, while a more complex method involves multiple tanks. It is highly sustainable to use stored water in place of purified water for activities such as irrigation and flushing toilets. Additionally, using stored rainwater reduces the amount of runoff pollution, picked up from roofs and pavements that would normally enter streams through storm drains. The following equation can be used to estimate annual water supply:

Collection area (square feet) × Rainfall (inch/year) / 12 (inch/foot) = Cubic Feet of Water/Year

Cubic Feet/Year × 7.43 (Gallons/Cubic Foot) = Gallons/year

Note, however, this calculation does not account for losses such as evaporation or leakage.

Greywater systems function in sequestering used indoor water, such as laundry, bath and sink water, and filtering it for reuse. Greywater can be reused in irrigation and toilet flushing. There are two types of greywater systems: gravity fed manual systems and package systems. The manual systems do not require electricity but may require a larger yard space. The package systems require electricity but are self-contained and can be installed indoors.

Waste

As populations and resource demands climb, waste production contributes to emissions of carbon dioxide, leaching of hazardous materials into the soil and waterways, and methane emissions. In America alone, over the course of a decade, 500 trillion pounds of American resources will have been transformed into nonproductive wastes and gases. Thus, a crucial component of sustainable living is being waste conscious. One can do this by reducing waste, reusing commodities, and recycling.

There are a number of ways to reduce waste in sustainable living. Two methods to reduce paper waste are canceling junk mail like credit card and insurance offers and direct mail marketing and changing monthly paper statements to paperless emails. Junk mail alone accounted for 1.72 million tons of landfill waste in 2009. Another method to reduce waste is to buy in bulk, reducing packaging materials. Preventing food waste can limit the amount of organic waste sent to landfills producing the powerful greenhouse gas methane. Another example of waste reduction involves being cognizant of purchasing excessive amounts when buying materials with limited use like cans of paint. Non-hazardous or less hazardous alternatives can also limit the toxicity of waste.

By reusing materials, one lives more sustainably by not contributing to the addition of waste to landfills. Reusing saves natural resources by decreasing the necessity of raw material extraction. For example, reusable bags can reduce the amount of waste created by grocery shopping eliminating the need to create and ship plastic bags and the need to manage their disposal and recycling or polluting effects.

Recycling, a process that breaks down used items into raw materials to make new materials, is a particularly useful means of contributing to the renewal of goods. Recycling incorporates three primary processes; collection and processing, manufacturing, and purchasing recycled products. A natural example of recycling involves using food waste as compost to enrich the quality of soil, which can be carried out at home or locally with community composting. An offshoot of recycling, upcycling, strives to convert material into something of similar or greater value in its second life. By integrating measures of reusing, reducing, and recycling one can effectively reduce personal waste and use materials in a more sustainable manner.

SUSTAINABLE FOREST MANAGEMENT

Sustainable forest management is the management of forests according to the principles of sustainable development. Sustainable forest management has to keep the balance between three main pillars: ecological, economic and socio-cultural. Successfully achieving sustainable forest management will provide integrated benefits to all, ranging from safeguarding local livelihoods to protecting the biodiversity and ecosystems provided by forests, reducing rural poverty and mitigating some of the effects of climate change.

The "Forest Principles" adopted at The United Nations Conference on Environment and Development (UNCED) in Rio de Janeiro in 1992 captured the general international understanding of sustainable forest management at that time. A number of sets of criteria and indicators have since been developed to evaluate the achievement of SFM at the global, regional, country and management unit level. These were all attempts to codify and provide for independent assessment of the degree to which the broader objectives of sustainable forest management are being achieved in practice. In 2007, the United Nations General Assembly adopted the Non-Legally Binding Instrument on All Types of Forests. The instrument was the first of its kind, and reflected the strong international commitment to promote implementation of sustainable forest management through a new approach that brings all stakeholders together.

A definition of SFM was developed by the Ministerial Conference on the Protection of Forests in Europe (FOREST EUROPE), and has since been adopted by the Food and Agriculture Organization (FAO). It defines sustainable forest management as:

> The stewardship and use of forests and forest lands in a way, and at a rate, that maintains their biodiversity, productivity, regeneration capacity, vitality and their potential to fulfill, now and in the future, relevant ecological, economic and social functions, at local, national, and global levels, and that does not cause damage to other ecosystems.

In simpler terms, the concept can be described as the attainment of balance – balance between society's increasing demands for forest products and benefits, and the preservation of forest health and diversity. This balance is critical to the survival of forests, and to the prosperity of forest-dependent communities.

For forest managers, sustainably managing a particular forest tract means determining, in a tangible way, how to use it today to ensure similar benefits, health and productivity in the future. Forest managers must assess and integrate a wide array of sometimes conflicting factors – commercial and non-commercial values, environmental considerations, community needs, even global impact – to produce sound forest plans. In most cases, forest managers develop their forest plans in consultation with citizens, businesses, organizations and other interested parties in and around the forest tract being managed. The tools and visualization have been recently evolving for better management practices.

The Food and Agriculture Organization of the United Nations, at the request of Member States, developed and launched the Sustainable Forest Management Toolbox in 2014, an online collection of tools, best practices and examples of their application to support countries implementing sustainable forest management.

Because forests and societies are in constant flux, the desired outcome of sustainable forest management is not a fixed one. What constitutes a sustainably managed forest will change over time as values held by the public change.

Criteria and Indicators

Criteria and indicators are tools which can be used to conceptualise, evaluate and implement sustainable forest management. Criteria define and characterize the essential elements, as well as a set of conditions or processes, by which sustainable forest management may be assessed. Periodically measured indicators reveal the direction of change with respect to each criterion.

Deforestation of native rain forest in Rio de Janeiro City for extraction of clay for civil engineering.

Criteria and indicators of sustainable forest management are widely used and many countries produce national reports that assess their progress toward sustainable forest management. There are nine international and regional criteria and indicators initiatives, which collectively involve more than 150 countries. Three of the more advanced initiatives are those of the Working Group on Criteria and Indicators for the Conservation and Sustainable Management of Temperate and Boreal Forests (also called the Montréal Process), Forest Europe, and the International Tropical Timber Organization. Countries who are members of the same initiative usually agree to produce reports at the same time and using the same indicators. Within countries, at the management unit level, efforts have also been directed at developing local level criteria and indicators of sustainable forest management. The Center for International Forestry Research, the International Model Forest Network and researchers at the University of British Columbia have developed a number of tools and techniques to help forest-dependent communities develop their own local level criteria and indicators. Criteria and Indicators also form the basis of third-party forest certification programs such as the Canadian Standards Association's Sustainable Forest Management Standards and the Sustainable Forestry Initiative Standard.

There appears to be growing international consensus on the key elements of sustainable forest management. Seven common thematic areas of sustainable forest management have emerged based on the criteria of the nine ongoing regional and international criteria and indicators initiatives. The seven thematic areas are:

- Extent of forest resources,

- Biological diversity,

- Forest health and vitality,

- Productive functions and forest resources,

- Protective functions of forest resources,

- Socio-economic functions,

- Legal, policy and institutional framework.

This consensus on common thematic areas (or criteria) effectively provides a common, implicit definition of sustainable forest management. The seven thematic areas were acknowledged by the international forest community at the fourth session of the United Nations Forum on Forests and the 16th session of the Committee on Forestry. These thematic areas have since been enshrined in the Non-Legally Binding Instrument on All Types of Forests as a reference framework for sustainable forest management to help achieve the purpose of the instrument.

On January 5, 2012, the Montréal Process, Forest Europe, the International Tropical Timber Organization, and the Food and Agriculture Organization of the United Nations, acknowledging the seven thematic areas, endorsed a joint statement of collaboration to improve global forest related data collection and reporting and avoiding the proliferation of monitoring requirements and associated reporting burdens.

Ecosystem Approach

The Ecosystem Approach has been prominent on the agenda of the Convention on Biological Diversity (CBD) since 1995. The CBD definition of the Ecosystem Approach and a set of principles for its application were developed at an expert meeting in Malawi in 1995, known as the Malawi Principles. The definition, 12 principles and 5 points of "operational guidance" were adopted by the fifth Conference of Parties (COP5) in 2000. The CBD definition is as follows:

> The ecosystem approach is a strategy for the integrated management of land, water and living resources that promotes conservation and sustainable use in an equitable way. Application of the ecosystem approach will help to reach a balance of the three objectives of the Convention. An ecosystem approach is based on the application of appropriate scientific methodologies focused on levels of biological organization, which encompasses the essential structures, processes, functions and interactions among organisms and their environment. It recognizes that humans, with their cultural diversity, are an integral component of many ecosystems.

Sustainable forest management was recognized by parties to the Convention on Biological Diversity in 2004 to be a concrete means of applying the Ecosystem Approach to forest ecosystems. The two concepts, sustainable forest management and the ecosystem approach, aim at promoting conservation and management practices which are environmentally, socially and economically sustainable, and which generate and maintain benefits for both present and future generations. In Europe, the MCPFE and the Council for the Pan-European Biological and Landscape Diversity Strategy (PEBLDS) jointly recognized sustainable forest management to be consistent with the Ecosystem Approach in 2006.

Independent Certification

Growing environmental awareness and consumer demand for more socially responsible businesses helped third-party forest certification emerge in the 1990s as a credible tool for communicating the environmental and social performance of forest operations.

There are many potential users of certification, including: forest managers, scientists, policy makers, investors, environmental advocates, business consumers of wood and paper, and individuals.

With third-party forest certification, an independent organization develops standards of good forest management, and independent auditors issue certificates to forest operations that comply with those standards. Forest certification verifies that forests are well-managed – as defined by a particular standard – and chain-of-custody certification tracks wood and paper products from the certified forest through processing to the point of sale.

This rise of certification led to the emergence of several different systems throughout the world. As a result, there is no single accepted forest management standard worldwide, and each system takes a somewhat different approach in defining standards for sustainable forest management.

In its 2009–2010 Forest Products Annual Market Review United Nations Economic Commission for Europe/Food and Agriculture Organization stated: "Over the years, many of the issues that previously divided the (certification) systems have become much less distinct. The largest certification systems now generally have the same structural programmatic requirements."

Third-party forest certification is an important tool for those seeking to ensure that the paper and wood products they purchase and use come from forests that are well-managed and legally harvested. Incorporating third-party certification into forest product procurement practices can be a centerpiece for comprehensive wood and paper policies that include factors such as the protection of sensitive forest values, thoughtful material selection and efficient use of products.

There are more than fifty certification standards worldwide, addressing the diversity of forest types and tenures. Globally, the two largest umbrella certification programs are:

- Programme for the Endorsement of Forest Certification (PEFC).

- Forest Stewardship Council (FSC).

The area of forest certified worldwide is growing slowly. PEFC is the world's largest forest certification system, with more than two-thirds of the total global certified area certified to its Sustainability Benchmarks.

In North America, there are three certification standards endorsed by PEFC – the Sustainable Forestry Initiative, the Canadian Standards Association's Sustainable Forest Management Standard, and the American Tree Farm System. FSC has five standards in North America – one in the United States and four in Canada.

While certification is intended as a tool to enhance forest management practices throughout the world, to date most certified forestry operations are located in Europe and North America. A significant barrier for many forest managers in developing countries is that they lack the capacity to undergo a certification audit and maintain operations to a certification standard.

Forest Governance

Although a majority of forests continue to be owned formally by government, the effectiveness of forest governance is increasingly independent of formal ownership. Since neo-liberal ideology in the 1980s and the emanation of the climate change challenges, evidence that the state is failing to effectively manage environmental resources has emerged. Under neo-liberal regimes in the developing countries, the role of the state has diminished and the market forces have increasingly taken over the dominant socio-economic role. Though the critiques of neo-liberal policies have maintained that market forces are not only inappropriate for sustaining the environment, but are in fact a major cause of environmental destruction. Hardin's tragedy of the commons has shown that the people cannot be left to do as they wish with land or environmental resources. Thus, decentralization of management offers an alternative solution to forest governance.

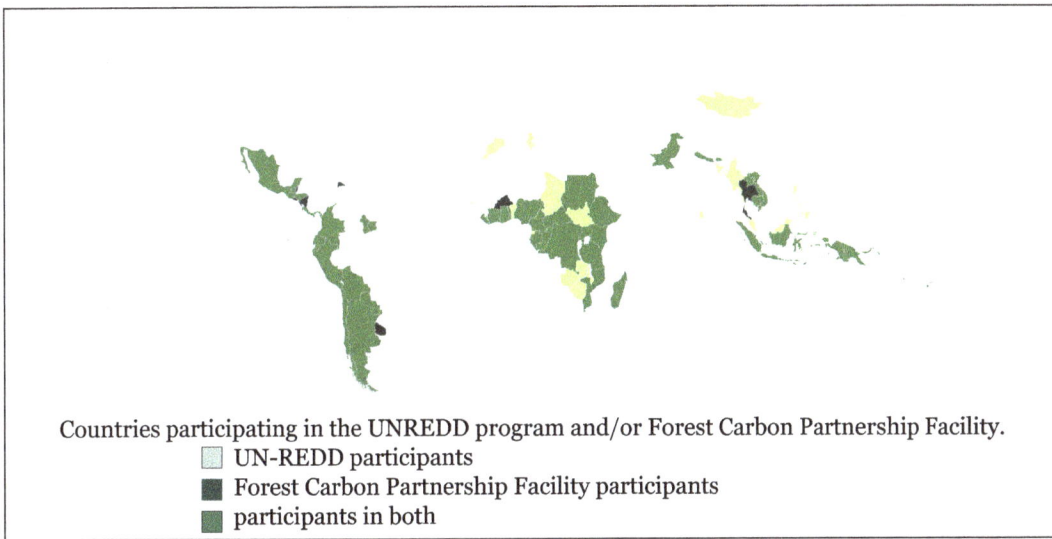

Countries participating in the UNREDD program and/or Forest Carbon Partnership Facility.
☐ UN-REDD participants
■ Forest Carbon Partnership Facility participants
■ participants in both

The shifting of natural resource management responsibilities from central to state and local governments, where this is occurring, is usually a part of broader decentralization process. According to Rondinelli and Cheema, there are four distinct decentralization options: these are: (i) Privatization – the transfer of authority from the central government to non-governmental sectors otherwise known as market-based service provision, (ii) Delegation – centrally nominated local authority, (iii) Devolution – transfer of power to locally acceptable authority and (iv) Deconcentration – the redistribution of authority from the central government to field delegations of the central government. The major key to effective decentralization is increased broad-based participation in local-public decision making. In 2000, the World Bank report reveals that local government knows the needs and desires of their constituents better than the national government, while at the same time, it is easier to hold local leaders accountable. From the study of West African tropical forest, it is argued that the downwardly accountable and representative authorities with meaningful discretional powers are the basic institutional element of decentralization that should lead to efficiency, development and equity. This collaborates with the World Bank report in 2000 which says that decentralization should improve resource allocation, efficiency, accountability and equity "by linking the cost and benefit of local services more closely".

Many reasons point to the advocacy of decentralization of forest: (i) Integrated rural development projects often fail because they are top-down projects that did not take local people's needs and desires into account. (ii) National government sometimes have legal authority over vast forest areas that they cannot control, thus, many protected area projects result in increased biodiversity loss and greater social conflict. Within the sphere of forest management, as state earlier, the most effective option of decentralization is "devolution"-the transfer of power to locally accountable authority. However, apprehension about local governments is not unfounded. They are often short of resources, may be staffed by people with low education and are sometimes captured by local elites who promote clientelist relation rather than democratic participation. Enters and Anderson point that the result of community-based projects intended to reverse the problems of past central approaches to conservation and development have also been discouraging.

Broadly speaking, the goal of forest conservation has historically not been met when, in contrast with land use changes; driven by demand for food, fuel and profit. It is necessary to recognize and advocate for better forest governance more strongly given the importance of forest in meeting basic human needs in the future and maintaining ecosystem and biodiversity as well as addressing climate change mitigation and adaptation goal. Such advocacy must be coupled with financial incentives for government of developing countries and greater governance role for local government, civil society, private sector and NGOs on behalf of the "communities".

National Forest Funds

The development of National Forest Funds is one way to address the issue of financing sustainable forest management. National forest funds (NFFs) are dedicated financing mechanisms managed by public institutions designed to support the conservation and sustainable use of forest resources. As of 2014, there are 70 NFFs operating globally.

Forest Genetic Resources

Appropriate use and long-term conservation of forest genetic resources (FGR) is a part of sustainable forest management. In particular when it comes to the adaptation of forests and forest management to climate change. Genetic diversity ensures that forest trees can survive, adapt and evolve under changing environmental conditions. Genetic diversity in forests also contributes to tree vitality and to the resilience towards pests and diseases. Furthermore, FGR has a crucial role in maintaining forest biological diversity at both species and ecosystem levels.

Selecting carefully the forest reproductive material with emphasis on getting a high genetic diversity rather than aiming at producing a uniform stand of trees, is essential for sustainable use of FGR. Considering the provenance is crucial as well. For example in relation to climate change, local material may not have the genetic diversity or phenotypic plasticity to guarantee good performance under changed conditions. A different population from further away, which may have experienced selection under conditions more like those forecast for the site to be reforested, might represent a more suitable seed source.

Methods of Sustainable Forest Management

A range of forestry institutions now practice various forms of sustainable forest management and a broad range of methods and tools are available that have been tested over time. The stewardship and use of forests and forest lands in a way, and at a rate, that maintains their biodiversity, productivity, regeneration capacity, vitality and their potential to fulfill, now and in the future, relevant ecological, economic and social functions, at local, national, and global levels, and that does not cause damage to other ecosystems.

The Seven Thematic Areas are:

1. Extent of forest resources.

2. Biological diversity.

3. Forest health and vitality.

4. Productive functions and forest resources.

5. Protective functions of forest resources.

6. Socio-economic functions.

7. Legal, policy and institutional framework.

There appears to be growing international consensus on the key elements of sustainable forest management. Seven common thematic areas of sustainable forest management have emerged based on the criteria of the nine ongoing regional and international criteria and indicators initiatives.

For the conservation of forests, following steps can be taken:

1. Conservation of forest is a national problem so it must be tackled with perfect coordination between forest department and other departments.

2. People's participation in the conservation of forests is of vital importance. So, we must get them involved in this national task.

3. The cutting of trees in the forests must be stopped at all costs.

4. Celebrations of all functions, festivals should proceed with tree-plantation.

5. Cutting of timber and other forest produce should be restricted.

6. Grasslands should be regenerated.

7. Forest Conservation Act, 1980 should be strictly implemented to check deforestation.

References

- Brinkmann, Svend (2012). Qualitative Inquiry in Everyday Life: Working with Everyday Life Materials. 1 Oliver's Yard, 55 City Road, London EC1Y 1SP United Kingdom: SAGE Publications Ltd. Doi:10.4135/9781473913905. ISBN 9780857024763

- Important-strategies-to-achieve-sustainable-development, essay: yourarticlelibrary.com, Retrieved 1 March, 2019

- Jonkers, Henk M. (23 March 2018). "Self Healing Concrete: A Biological Approach". Self Healing Materials. Springer Series in Materials Science. 100. Springer, Dordrecht. Pp. 195–204. Doi:10.1007/978-1-4020-6250-6-9. ISBN 978-1-4020-6249-0

- World Energy Council (2007). "Transport Technologies and Policy Scenarios". World Energy Council. Archived from the original on 2008-12-04. Retrieved 2009-05-26

- Sustainable-forest-management-and-forest-conservation-methods, forest, environment: yourarticlelibrary. com, Retrieved 2 April, 2019

The technologies which focus on reducing waste while maintaining efficiency as well as improving the social and environmental footprint during the production cycle are known as sustainable technologies. The diverse aspects of sustainable technologies and sustainable energy have been thoroughly discussed in this chapter.

Sustainable technology relies on resources that are either renewable or so abundant that we can treat them as such. For technology to be sustainable also means that using it does not have any long-term adverse impact on the environment.

Sustainable technologies have become a vital factor across a wide span of industries. Why? There are some important reasons why sustainable technologies have become such a big deal.

Enduring Supply

One of the key ideas behind sustainable technologies and the philosophy of sustainability in general is creating an enduring supply. Primarily, this refers to the supply of materials that go into manufacturing, whether it's what is used to make products, or what is used in the process.

Traditional methods of obtaining or creating these elements, such as mining, are tapping into a finite resource. Sustainability means finding a way to replace a resource as it is used, creating an ecosystem that avoids the elimination of natural supplies or the surrounding environment.

Conserving the Environment

It is difficult to deny the overall impact industrial development has had on the global environment. From the destruction of rainforests to the pollution of the atmosphere, the byproducts of industrial activity over time have been intensely harmful.

While sustainable technologies and manufacturing processes are not necessarily a magic bullet solution, the focus on minimizing the damage and creating a sustainable supply source makes a real difference. There is a reason that so many companies have opened up to green technology solutions — they work.

Taking Responsibility

Taking responsibility for the impact technology and industry can have on society and the environment can be difficult, but it is vital. Education is the key here, and teaching and promoting sustainability is an important part of a sustainability-focused ethos.

Lasting Opportunities for Innovation

A side benefit of creating a sustainable, eco-friendly supply of materials is the creation of a lasting resource for innovation. With a sustainable supply, creative innovators in a wide variety of industries have more opportunity to invent new products, processes or methods of manufacturing. This means that, ultimately, the benefits of sustainability go beyond its positive impact on our environment.

Green Technology

As the world of technology progresses, it becomes more and more integrated into our everyday lives. As such, many people and businesses use various forms of technology every day. This constant usage has driven a need for sustainability and greener technologies. Although technology is amazing and helpful, if it's doing harm to the environment in the long term, it becomes detrimental instead of useful. That's why companies create green technology solutions to solve this potential world-changing problem.

Green technology is technology that is designed to be sustainable and environmentally friendly. Sustainable means that something is designed to meet a need without depleting natural resources, so that future generations are able to meet those same needs. Environmentally friendly could mean a number of things. It could mean reducing pollutants, conserving water usage, or saving on energy consumption. It could also mean creating something that can be reused or recycled, instead of ending up in a landfill. Brightsync is an example of a green technology that helps to drastically cut down on energy consumption for fleets.

Use of Green Technology

When it comes to the question of whether or not green technology is right for you or your business, there are two major benefits to consider. The first and most obvious benefit is the reduction of your environmental footprint. By using green technology, you're helping to make the world a more sustainable and better place to live. The other main draw for businesses is the amount of money they can save. Green technologies are often designed in a way that can save companies a large amount of money in utilities or waste removal. Plus, there are often incentive programs and rebates that companies can get from implementing green technology in their business model. It's a win-win situation.

SUSTAINABLE ENERGY

Sustainable energy is a form of energy that meet our today's demand of energy without putting them in danger of getting expired or depleted and can be used over and over again. Sustainable energy should be widely encouraged as it do not cause any harm to the environment and is available widely free of cost. All renewable energy sources like solar, wind, geothermal, hydropower and ocean energy are sustainable as they are stable and available in plenty.

Sun will continue to provide sunlight till we all are here on earth, heat caused by sun will continue to produce winds, earth will continue to produce heat from inside and will not cool down anytime soon, movement of earth, sun and moon will not stop and this will keep on producing tides and the process of evaporation will cause water to evaporate that will fall down in the form of rain or ice which will go through rivers or streams and merge in the oceans and can be used to produce energy through hydropower. This clearly states that all these renewable energy sources are sustainable and will continue to provide energy to the coming generations.

There are many forms of sustainable energy sources that can be incorporated by countries to stop the use of fossil fuels. Sustainable energy does not include any sources that are derived from fossil fuels or waste products. This energy is replenishable and helps us to reduce greenhouse gas emissions and causes no damage to the environment. If we are going to use fossil fuels at a steady rate, they will expire soon and cause adverse affect to our planet.

Fossil fuels are not considered as sustainable energy sources because they are limited, cause immense pollution by releasing harmful gases and are not available everywhere on earth. Fossil fuels normally include coal, oil and natural gas. Steps must be taken to reduce our dependency on fossil fuels as pose dangerous to environment. Most of the counties have already started taking steps to make use of alternative energy sources. As of today, around 20% of world's energy needs comes from renewable energy sources. Hydropower is the most common form of alternative energy used around the world.

Need for Sustainable Energy

During ancient times, wood, timber and waste products were the only major energy sources. In short, biomass was the only way to get energy. When more technology was developed, fossil fuels like coal, oil and natural gas were discovered. Fossil fuels proved boom to the mankind as they were widely available and could be harnessed easily. When these fossil fuels were started using extensively by all the countries across the globe, they led to degradation of environment. Coal and oil are two of the major sources that produce large amount of carbon dioxide in the air. This led to increase in global warming.

Also, few countries have hold on these valuable products which led to the rise in prices of these fuels. Now, with rising prices, increasing air pollution and risk of getting expired soon forced scientists to look out for some alternative or renewable energy sources. The need of the hour was to look for resources that are available widely, cause no pollution and are replenishable. Sustainable Energy, at that time came into the picture as it could meet our today's increasing demand of energy and also provide us with an option to make use of them in future also.

Renewable Energy Sources

When referring to sources of energy, the terms "sustainable energy" and "renewable energy" are often used interchangeably, however particular renewable energy projects sometimes raise significant sustainability concerns. Renewable energy technologies are essential contributors to sustainable energy as they generally contribute to world energy security, reducing dependence on fossil fuel resources, and providing opportunities for mitigating greenhouse gases. Various Cost–benefit analysis work by a disparate array of specialists and agencies have been conducted to determine the cheapest and quickest paths to decarbonizing the energy supply of the world, with the topic being one of considerable controversy, particularly on the role of nuclear energy.

Hydropower

Among sources of renewable energy, hydroelectric plants have the advantages of being long-lived—many existing plants have operated for more than 100 years. Also, hydroelectric plants are clean and have few emissions. Criticisms directed at large-scale hydroelectric plants include: dislocation of people living where the reservoirs are planned, and release of significant amounts of carbon dioxide during construction and flooding of the reservoir.

Hydroelectric dams are one of the most widely deployed sources of sustainable energy.

However, it has been found that high emissions are associated only with shallow reservoirs in warm (tropical) locales, and recent innovations in hydropower turbine technology are enabling efficient development of low-impact run-of-the-river hydroelectricity projects. Generally speaking, hydroelectric plants produce much lower life-cycle emissions than other types of generation. Hydroelectric power, which underwent extensive development during growth of electrification in the 19th and 20th centuries, is experiencing resurgence of development in the 21st century. The areas of greatest hydroelectric growth are the booming economies of Asia. China is the development leader; however, other Asian nations are installing hydropower at a rapid pace. This growth is driven by much increased energy costs—especially for imported energy—and widespread desires for more domestically produced, clean, renewable, and economical generation.

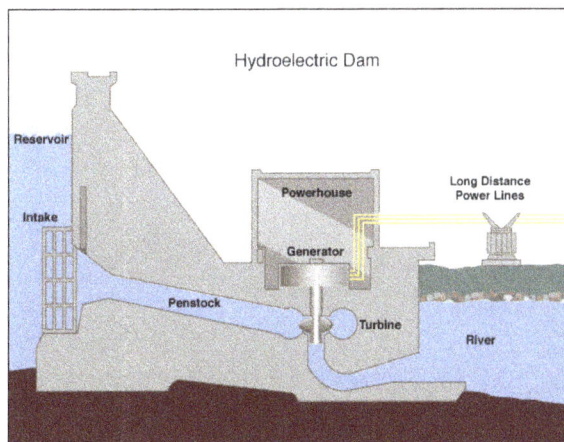

Hydroelectric dam in cross section.

Geothermal

Geothermal energy can be harnessed to for electricity generation and for heating. Technologies in use include dry steam power stations, flash steam power stations and binary cycle power stations. As of 2010, geothermal electricity generation is used in 24 countries, while geothermal heating is in use in 70 countries. International markets grew at an average annual rate of 5 percent over the three years to 2015, and global geothermal power capacity is expected to reach 14.5–17.6 GW by 2020.

One of many power plants at The Geysers, a geothermal power field in northern California, with a total output of over 750 MW.

Geothermal power is considered to be a sustainable, renewable source of energy because the heat extraction is small compared with the Earth's heat content. The greenhouse gas emissions of geothermal electric stations are on average 45 grams of carbon dioxide per kilowatt-hour of electricity, or less than 5 percent of that of conventional coal-fired plants. As a source of renewable energy for both power and heating, geothermal has the potential to meet 3-5% of global demand by 2050. With economic incentives, it is estimated that by 2100 it will be possible to meet 10% of global demand.

Biomass and Biofuel

Sugarcane plantation to produce ethanol in Brazil.

A CHP power station using wood to provide electricity to over 30.000 households in France.

Biomass is biological material derived from living, or recently living organisms. As an energy source, biomass can either be burned to produce heat and to generate electricity, or converted to various forms of biofuel. Liquid biofuels such as biodiesel and ethanol are especially valued as energy sources for motor vehicles.

Biomass is extremely versatile and one of the most-used sources of renewable energy. It is available in many countries, which makes it attractive for reducing dependence on imported fossil fuels. If the production of biomass is well-managed, carbon emissions can be significantly offset by the absorption of carbon dioxide by the plants during their lifespans. If the biomass source is agricultural or municipal waste, burning it or converting it into biogas also provides a way to dispose of this waste.

As of 2012, wood remains the largest biomass energy source today. If biomass is harvested from crops, such as tree plantations, the cultivation of these crops can displace natural ecosystems, degrade soils, and consume water resources and synthetic fertilizers. In some cases, these impacts can actually result in higher overall carbon emissions compared to using petroleum-based fuels.

Use of farmland for growing fuel can result in less land being available for growing food. Since photosynthesis is inherently inefficient, and crops also require significant amounts of energy to harvest, dry, and transport, the amount of energy produced per unit of land area is very small, in the range of 0.25 W/m² to 1.2 W/m². In the United States, corn-based ethanol has replaced less than 10% of motor gasoline use since 2011, but has consumed around 40% of the annual corn harvest in the country. In Malaysia and Indonesia, the clearing of forests to produce palm oil for biodiesel has led to serious social and environmental effects, as these forests are critical carbon sinks and habitats for endangered species.

Wind

In Europe in the 19th century, there were about 200,000 windmills, slightly more than the modern wind turbines of the 21st century. They were mainly used to grind grain and to pump water. The age of coal powered steam engines replaced this early use of wind power.

Wind power has high potential and have already realised relatively low production costs. At the end of 2008, worldwide wind farm capacity was 120,791 megawatts (MW), representing an increase of

28.8 percent during the year, and wind power produced some 1.3% of global electricity consumption. Wind power accounts for approximately 20% of electricity use in Denmark, 9% in Spain, and 7% in Germany. However, it may be difficult to site wind turbines in some areas for aesthetic or environmental reasons, and it may be difficult to integrate wind power into electricity grids in some cases.

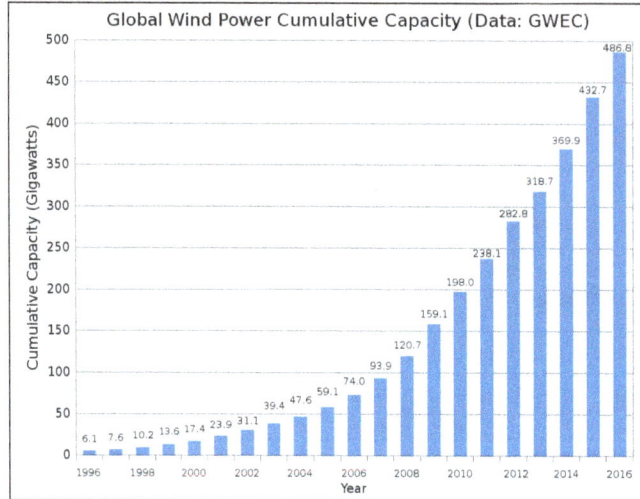

Wind power: worldwide installed capacity.

Solar Heating

Solar heating systems generally consist of solar thermal collectors, a fluid system to move the heat from the collector to its point of usage, and a reservoir or tank for heat storage and subsequent use. The systems may be used to heat domestic hot water, swimming pool water, or for space heating. The heat can also be used for industrial applications or as an energy input for other uses such as cooling equipment. In many climates, a solar heating system can provide a very high percentage (20 to 80%) of domestic hot water energy. Energy received from the sun by the earth is that of electromagnetic radiation. Light ranges of visible, infrared, ultraviolet, x-rays, and radio waves received by the earth through solar energy. The highest power of radiation comes from visible light. Solar power is complicated due to changes in seasons and from day to night. Cloud cover can also add to complications of solar energy, and not all radiation from the sun reaches earth because it is absorbed and dispersed due to clouds and gases within the earth's atmospheres.

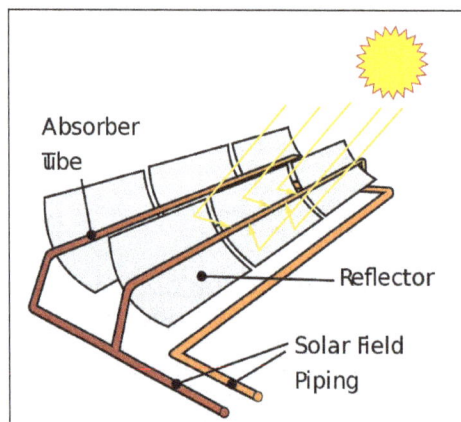

Sketch of a Parabolic Trough Collector.

Solar thermal power stations have been successfully operating in California commercially since the late 1980s, including the largest solar power plant of any kind, the 350 MW Solar Energy Generating Systems. Nevada Solar One is another 64MW plant which has recently opened. Other parabolic trough power plants being proposed are two 50 MW plants in Spain, and a 100 MW plant in Israel.

Solar Electricity

Solar electricity production uses photovoltaic (PV) cells to convert light into electrical current. Photovoltaic modules can be integrated into buildings or used in photovoltaic power stations connected to the electrical grid. They are especially useful for providing electricity to remote areas.

11 MW solar power plant near Serpa, Portugal.

Large national and regional research projects on artificial photosynthesis are designing nanotechnology-based systems that use solar energy to split water into hydrogen fuel and a proposal has been made for a Global Artificial Photosynthesis project. In 2011, researchers at the Massachusetts Institute of Technology (MIT) developed what they are calling an "artificial leaf", which is capable of splitting water into hydrogen and oxygen directly from solar power when dropped into a glass of water. One side of the "Artificial Leaf" produces bubbles of hydrogen, while the other side produces bubbles of oxygen.

MIT's Solar House built in 1939 used seasonal thermal energy storage (STES) for year-round heating.

Most current solar power plants are made from an array of similar units where each unit is continuously adjusted, e.g., with some step motors, so that the light converter stays in focus of the sun light. The cost of focusing light on converters such as high-power solar panels, stirling engine, etc. can be dramatically decreased with a simple and efficient rope mechanics. In this technique many units are connected with a network of ropes so that pulling two or three ropes is sufficient to keep all light converters simultaneously in focus as the direction of the sun changes.

Research is ongoing in space-based solar power, a concept in which solar panels are launched into outer space and the energy they capture is transmitted back to Earth as microwaves. A test facility for the technology is being built in China.

Ocean Energy

The world's first commercial tidal stream generator – SeaGen – in Strangford Lough.
The strong wake shows the power in the tidal current.

Portugal has the world's first commercial wave farm, the Aguçadora Wave Park, under construction in 2007. The farm will initially use three Pelamis P-750 machines generating 2.25 MW and costs are put at 8.5 million euro. Subject to successful operation, a further 70 million euro is likely to be invested before 2009 on a further 28 machines to generate 525 MW. Funding for a wave farm in Scotland was announced in February, 2007 by the Scottish Executive, at a cost of over 4 million pounds, as part of a £13 million funding packages for ocean power in Scotland. The farm will be the world's largest with a capacity of 3 MW generated by four Pelamis machines.

In 2007, the world's first turbine to create commercial amounts of energy using tidal power was installed in the narrows of Strangford Lough in Northern Ireland, UK. The 1.2 MW underwater tidal electricity generator takes advantage of the fast tidal flow in the lough which can be up to 4 m/s. Although the generator is powerful enough to power up to a thousand homes, the turbine has a minimal environmental impact, as it is almost entirely submerged, and the rotors turn slowly enough that they pose no danger to wildlife.

Enabling Technologies for Variable Renewable Energy

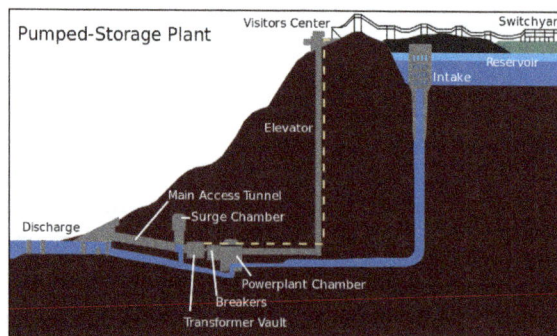

In a pumped-storage hydroelectricity facility, water is pumped uphill electricity generation exceeds demand. The water is later released to generate hydroelectricity.

Solar and wind are intermittent energy sources that supply electricity 10-40% of the time, depending on the weather and the time of day. Most electric grids were constructed for non-intermittent energy sources such as hydroelectricity or coal-fired power plants. In general, up to around 30% of the energy supplied to an electric grid can be easily converted to intermittent sources.

If intermittent sources make up a larger percentage of the energy supply for a given electric grid, there are several possible approaches to ensuring that electricity generation can meet ongoing demand:

- Reducing demand for electricity at certain times through energy demand management and use of smart grids.

- Using hydroelectricity or natural gas generation to produce backup power.

- Importing electricity from other locations through long-distance transmission lines. For example, TREC has proposed to distribute solar power from the Sahara to Europe. Europe can distribute wind and ocean power to the Sahara and other countries. In this way, power is produced at any given time as at any point of the planet as the sun or the wind is up or ocean waves and currents are stirring.

- Using grid energy storage to store excess solar and wind energy and release it as needed. The most commonly-used storage method is pumped-storage hydroelectricity, which is feasible only at locations that are next to a large hill or a deep underground mine. Other storage technologies are flywheel energy storage, compressed air, batteries, and hydrogen fuel.

As of 2019, the cost and logistics of energy storage for large population centres is a significant challenge, although the cost of battery systems has plunged dramatically. For instance, a 2019 study found that for solar and wind energy to meet energy demand for a week of extreme cold in the eastern and midwest United States, energy storage capacity would have to increase from the 11 GW currently in place to 277.9 GW.

Some costs could potentially be reduced by making use of energy storage equipment the consumer buys and not the state. An example is batteries in electric cars that would double as an energy buffer for the electricity grid. Energy storage apparatus' as car batteries are also built with materials that pose a threat to the environment (e.g. Lithium). The combined production of batteries for such a large part of the population would still have environmental concerns.

To provide household electricity in remote areas (that is areas which are not connected to the mains electricity grid), energy storage is required for use with renewable energy. Energy generation and consumption systems used in the latter case are usually stand-alone power systems.

Energy from renewable sources can also be stored as heat or cold, through thermal energy storage technologies. For instance, summer heat can be stored for winter heating, or winter cold can be stored for summer air conditioning.

Non-renewable Energy Sources

There is considerable controversy over whether nuclear power can be considered sustainable. Some forms of nuclear power (ones which are able to "burn" nuclear waste through a process known as nuclear transmutation, such as an Integral Fast Reactor, could belong in the "Green Energy" category). Nuclear power plants can be more or less eliminated from their problem of nuclear waste through the use of nuclear reprocessing and newer plants as fast breeder and nuclear fusion plants.

Some people, including early Greenpeace member Patrick Moore, George Monbiot, Bill Gates and James Lovelock have specifically classified nuclear power as green energy. Others, including Greenpeace's Phil Radford disagree, claiming that the problems associated with radioactive waste and the risk of nuclear accidents (such as the Chernobyl disaster) pose an unacceptable risk to the environment and to humanity. However, newer nuclear reactor designs are capable of utilizing what is now deemed "nuclear waste" until it is no longer (or dramatically less) dangerous, and have design features that greatly minimize the possibility of a nuclear accident. These designs have yet to be commercialized.

In theory, the greenhouse gas emissions of fossil fuel power plants can be significantly reduced through carbon capture and storage, although this process is expensive. Some believe that fossil fuel burning, with carbon capture and storage, may have a role in a sustainable energy system.

Energy Efficiency

Moving towards energy sustainability will require changes not only in the way energy is supplied, but in the way it is used, and reducing the amount of energy required to deliver various goods or services is essential. Opportunities for improvement on the demand side of the energy equation are as rich and diverse as those on the supply side, and often offer significant economic benefits.

Efficiency slows down energy demand growth so that rising clean energy supplies can make deep cuts in fossil fuel use. A recent historical analysis has demonstrated that the rate of energy efficiency improvements has generally been outpaced by the rate of growth in energy demand, which is due to continuing economic and population growth. As a result, despite energy efficiency gains, total energy use and related carbon emissions have continued to increase. Thus, given the thermodynamic and practical limits of energy efficiency improvements, slowing the growth in energy demand is essential. However, unless clean energy supplies come online rapidly, slowing demand growth will only begin to reduce total emissions; reducing the carbon content of energy sources is also needed. Any serious vision of a sustainable energy economy thus requires commitments to both renewables and efficiency.

Clean Cookstoves

In developing countries, an estimated 3 billion people rely on traditional cookstoves and open fires to burn biomass or coal for heating and cooking. This practice causes harmful local air pollution and increases danger from fires, resulting in an estimated 4.3 million deaths annually. Additionally, serious local environmental damage, including desertification, can be caused by excessive harvesting of wood and other combustible material. Promoting usage of cleaner fuels and more efficient technologies for cooking is therefore one of the top priorities of the United Nations Sustainable Energy for All initiative. Thus far, efforts to design cookstoves that are inexpensive, powered by sustainable energy sources, and acceptable to users have been mostly disappointing.

Trends

Climate change concerns coupled with high oil prices and increasing government support are driving increasing rates of investment in the sustainable energy industries, according to a trend analysis from the United Nations Environment Programme. According to UNEP, global investment in sustainable energy in 2007 was higher than previous levels, with $148 billion of new money raised in 2007, an increase of 60% over 2006. Total financial transactions in sustainable energy, including acquisition activity, was $204 billion.

Investment flows in 2007 broadened and diversified, making the overall picture one of greater breadth and depth of sustainable energy use. The mainstream capital markets are "now fully receptive to sustainable energy companies, supported by a surge in funds destined for clean energy investment". The increased levels of investment and the fact that much of the capital is coming from more conventional financial actors suggest that sustainable energy options are now becoming mainstream.

Purchasing Green Electricity

In several countries with common carrier arrangements, electricity retailing arrangements make it possible for consumers to purchase "green" electricity from either their utility or a green power provider. Electricity is considered to be green if it is produced from a source that produces relatively little pollution, and the concept is often considered equivalent to renewable energy.

A solar trough array is an example of green energy.

Public seat with integrated solar panel in Singapore-anyone can sit and plug in their mobile phone for free charging.

In many countries, green energy currently provides a very small amount of electricity, generally contributing less than 2 to 5% to the overall pool of electricity offered by most utility companies, electric companies, or state power pools. In some U.S. states, local governments have formed regional power purchasing pools using Community Choice Aggregation and Solar Bonds to achieve a 51% renewable mix or higher, such as in the City of San Francisco.

By participating in a green energy program a consumer may be having an effect on the energy sources used and ultimately might be helping to promote and expand the use of green energy. They are also making a statement to policy makers that they are willing to pay a price premium to support renewable energy. Green energy consumers either obligate the utility companies to increase the amount of green energy that they purchase from the pool (so decreasing the amount of non-green energy they purchase), or directly fund the green energy through a green power provider. If insufficient green energy sources are available, the utility must develop new ones or contract with a third party energy supplier to provide green energy, causing more to be built. However, there is no way the consumer can check whether or not the electricity bought is "green" or otherwise.

In some countries such as the Netherlands, electricity companies guarantee to buy an equal amount of 'green power' as is being used by their green power customers. The Dutch government exempts green power from pollution taxes, which means green power is hardly any more expensive than other power.

Green Energy and Labeling by Region

European Union

Directive 2004/8/EC of the European Parliament and of the Council of 11 February 2004 on the promotion of cogeneration based on a useful heat demand in the internal energy market includes the article 5 (Guarantee of origin of electricity from high-efficiency cogeneration).

European environmental NGOs have launched an ecolabel for green power. The ecolabel is called EKOenergy. It sets criteria for sustainability, additionality, consumer information and tracking. Only part of electricity produced by renewables fulfills the EKOenergy criteria.

A Green Energy Supply Certification Scheme was launched in the United Kingdom in February 2010. This implements guidelines from the Energy Regulator, Ofgem, and sets requirements on transparency, the matching of sales by renewable energy supplies, and additionality.

United States

The United States Department of Energy (DOE), the Environmental Protection Agency (EPA), and the Center for Resource Solutions (CRS) recognizes the voluntary purchase of electricity from renewable energy sources (also called renewable electricity or green electricity) as green power.

The most popular way to purchase renewable energy as revealed by NREL data is through purchasing Renewable Energy Certificates (RECs). According to a Natural Marketing Institute (NMI) survey 55 percent of American consumers want companies to increase their use of renewable energy.

DOE selected six companies for its 2007 Green Power Supplier Awards, including Constellation NewEnergy; 3Degrees; Sterling Planet; SunEdison; Pacific Power and Rocky Mountain Power; and Silicon Valley Power. The combined green power provided by those six winners equals more than 5 billion kilowatt-hours per year, which is enough to power nearly 465,000 average U.S. households. In 2014, Arcadia Power made RECS available to homes and businesses in all 50 states, allowing consumers to use "100% green power" as defined by the EPA's Green Power Partnership.

The U.S. Environmental Protection Agency (USEPA) Green Power Partnership is a voluntary program that supports the organizational procurement of renewable electricity by offering expert advice, technical support, tools and resources. This can help organizations lower the transaction costs of buying renewable power, reduce carbon footprint, and communicate its leadership to key stakeholders.

Throughout the country, more than half of all U.S. electricity customers now have an option to purchase some type of green power product from a retail electricity provider. Roughly one-quarter of the nation's utilities offer green power programs to customers, and voluntary retail sales of renewable energy in the United States totaled more than 12 billion kilowatt-hours in 2006, a 40% increase over the previous year.

In the United States, one of the main problems with purchasing green energy through the electrical grid is the current centralized infrastructure that supplies the consumer's electricity. This infrastructure has led to increasingly frequent brown outs and black outs, high CO_2 emissions, higher energy costs, and power quality issues. An additional $450 billion will be invested to expand this fledgling system over the next 20 years to meet increasing demand. In addition, this centralized system is now being further overtaxed with the incorporation of renewable energies such as wind, solar, and geothermal energies. Renewable resources, due to the amount of space they require, are often located in remote areas where there is a lower energy demand. The current infrastructure would make transporting this energy to high demand areas, such as urban centers, highly inefficient and in some cases impossible. In addition, despite the amount of renewable energy produced or the economic viability of such technologies only about 20 percent will be able to be incorporated into the grid. To have a more sustainable energy profile, the United States must move towards implementing changes to the electrical grid that will accommodate a mixed-fuel economy.

Several initiatives are being proposed to mitigate distribution problems. First and foremost, the most effective way to reduce USA's CO_2 emissions and slow global warming is through conservation efforts. Opponents of the current US electrical grid have also advocated for decentralizing the grid. This system would increase efficiency by reducing the amount of energy lost in transmission. It would also be economically viable as it would reduce the amount of power lines that will need to be constructed in the future to keep up with demand. Merging heat and power in this system would create added benefits and help to increase its efficiency by up to 80-90%. This is a significant increase from the current fossil fuel plants which only have an efficiency of 34%.

Small-scale Green Energy Systems

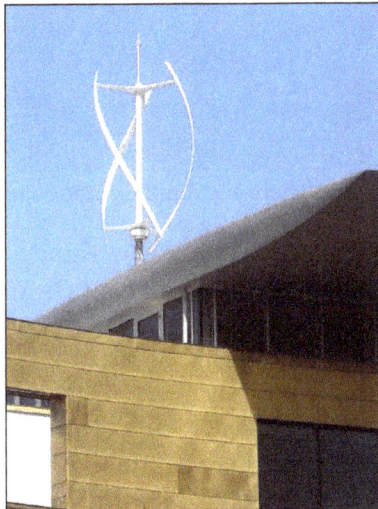

A small Quietrevolution QR5 Gorlov type vertical axis wind turbine in Bristol, England.
Measuring 3 m in diameter and 5 m high, it has a nameplate rating of 6.5 kW to the grid.

Those not satisfied with the third-party grid approach to green energy via the power grid can install their own locally based renewable energy system. Renewable energy electrical systems from solar to wind to even local hydro-power in some cases, are some of the many types of renewable energy systems available locally. Additionally, for those interested in heating and cooling their dwelling via renewable energy, geothermal heat pump systems that tap the constant temperature of the earth, which is around 7 to 15 degrees Celsius a few feet underground and increases dramatically at greater depths, are an option over conventional natural gas and petroleum-fueled heat approaches. Also, in geographic locations where the Earth's Crust is especially thin, or near volcanoes (as is the case in Iceland) there exists the potential to generate even more electricity than would be possible at other sites, thanks to a more significant temperature gradient at these locales.

The advantage of this approach in the United States is that many states offer incentives to offset the cost of installation of a renewable energy system. In California, Massachusetts and several other U.S. states, a new approach to community energy supply called Community Choice Aggregation has provided communities with the means to solicit a competitive electricity supplier and use municipal revenue bonds to finance development of local green energy resources. Individuals are usually assured that the electricity they are using is actually produced from a green energy source that they control. Once the system is paid for, the owner of a renewable energy system will be producing their own renewable electricity for essentially no cost and can sell the excess to the local utility at a profit.

A 01 KiloWatt Micro Windmill for Domestic Usage.

In household power systems, organic matter such as cow dung and spoilable organic matter can be converted to biochar. To eliminate emissions, carbon capture and storage is then used.

Sustainable Energy Research

There are numerous organizations within the academic, federal, and commercial sectors conducting large scale advanced research in the field of sustainable energy. Scientific production towards sustainable energy systems is rising exponentially, growing from about 500 English journal papers only about renewable energy in 1992 to almost 9,000 papers in 2011.

Biofuels

Cellulosic ethanol has many benefits over traditional corn based-ethanol. It does not take away or directly conflict with the food supply because it is produced from wood, grasses, or non-edible parts of plants. Moreover, some studies have shown cellulosic ethanol to be potentially more cost effective and economically sustainable than corn-based ethanol. As of 2018, efforts to commercialize production of cellulosic ethanol have been mostly disappointing, but new commercial efforts are continuing.

Algae fuel is an alternative to liquid fossil fuels that uses algae as its source of energy-rich oils. During the biofuel production process algae actually consumes the carbon dioxide in the air and turns it into oxygen through photosynthesis. In addition to its projected high yield, algaculture—unlike food crop-based biofuels — does not entail a decrease in food production, since it requires neither farmland nor fresh water. Between 2005 and 2012, dozens of companies attempted to commercialize production of algae fuel. By 2017, however, most efforts had been abandoned or changed to other applications, with only a few remaining.

Thorium

There are potentially two sources of nuclear power. Fission is used in all current nuclear power plants. Fusion is the reaction that exists in stars, including the sun, and remains impractical for use on Earth, as fusion reactors are not yet available. However, nuclear power is controversial politically and scientifically due to concerns about radioactive waste disposal, safety, the risks of a severe accident, and technical and economical problems in dismantling of old power plants.

Thorium is a fissionable material used in thorium-based nuclear power. The thorium fuel cycle claims several potential advantages over a uranium fuel cycle, including greater abundance, superior physical and nuclear properties, better resistance to nuclear weapons proliferation and reduced plutonium and actinide production. Therefore, it is sometimes referred as sustainable.

Solar

Currently, photovoltaic (PV) panels only have the ability to convert around 24% of the sunlight that hits them into electricity. At this rate, solar energy still holds many challenges for widespread implementation, but steady progress has been made in reducing manufacturing cost and increasing photovoltaic efficiency. In 2008, researchers at Massachusetts Institute of Technology (MIT) developed a method to store solar energy by using it to produce hydrogen fuel from water. Such research is targeted at addressing the obstacle that solar development faces of storing energy for use during nighttime hours when the sun is not shining. In February 2012, North Carolina-based Semprius Inc., announced that they had developed the world's most efficient solar panel. The company claims that the prototype converts 33.9% of the sunlight that hits it to electricity, more than double the previous high-end conversion rate. Major projects on artificial photosynthesis or solar fuels are also under way in many developed nations.

Wind

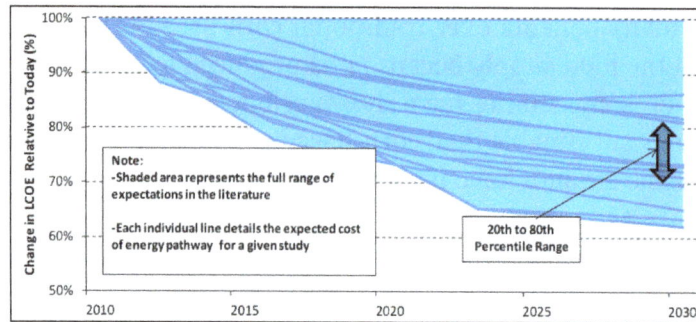

The National Renewable Energy Laboratory projects that the levelized cost of wind power in the U.S. will decline about 25% from 2012 to 2030.

Bangui Wind Farm in the Philippines.

Wind energy research dates back several decades to the 1970s when NASA developed an analytical model to predict wind turbine power generation during high winds. The Field Laboratory for Optimized Wind Energy (FLOWE) at Caltech was established to research renewable approaches to wind energy farming technology practices that have the potential to reduce the cost, size, and environmental impact of wind energy production. The president of Sky WindPower Corporation thinks that wind turbines will be able to produce electricity at a cent/kWh at an average which in comparison to coal-generated electricity is a fractional of the cost.

A wind farm is a group of wind turbines in the same location used to produce electric power. A large wind farm may consist of several hundred individual wind turbines, and cover an extended area of hundreds of square miles, but the land between the turbines may be used for agricultural or other purposes. A wind farm may also be located offshore.

Many of the largest operational onshore wind farms are located in the USA and China. Europe leads in the use of wind power with almost 66 GW, about 66 percent of the total globally, with Denmark in the lead according to the countries installed per-capita capacity.

Wind power has expanded quickly, its share of worldwide electricity usage at the end of 2014 was 3.1%.

Geothermal

Geothermal energy is produced by tapping into the thermal energy created and stored within the earth. It arises from the radioactive decay of an isotope of potassium and other elements found in the Earth's crust. Geothermal energy can be obtained by drilling into the ground, very similar to oil exploration, and then it is carried by a heat-transfer fluid (e.g. water, brine or steam). Geothermal systems that are mainly dominated by water have the potential to provide greater benefits to the system and will generate more power. Within these liquid-dominated systems, there are possible concerns of subsidence and contamination of ground-water resources. Therefore, protection of ground-water resources is necessary in these systems. This means that careful reservoir production and engineering is necessary in liquid-dominated geothermal reservoir systems. Geothermal energy is considered sustainable because that thermal energy is constantly replenished.

Hydrogen

Over $1 billion of federal money has been spent on the research and development of hydrogen and a medium for energy storage in the United States. Hydrogen is useful for energy storage, and for use in airplanes and ships, but is not practical for automobile use, as it is not very efficient, compared to using a battery — for the same cost a person can travel three times as far using a battery electric vehicle. Regardless of that opinion, Japanese car manufacturers Toyota and Honda currently offer hydrogen fuel-cell powered passenger vehicles for sale in Japan and the U.S.A. Experimental hydrogen fuel-cell city buses are currently operative in two U.S. transit districts, Alameda/Contra Costa county, California, and in Connecticut.

Government Promotion of Sustainable Energy

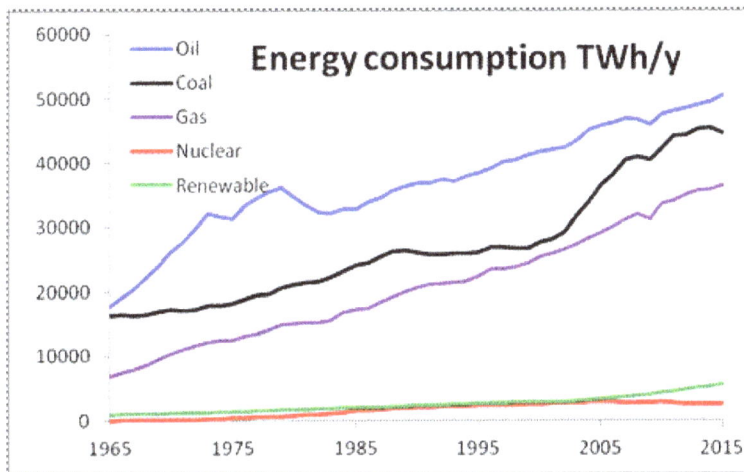

Comparing trends in worldwide energy use, the growth of clean energy to 2015 is shown by the green line.

Around the world many sub-national governments - regions, states and provinces - have aggressively pursued sustainable energy investments. In the United States, California's leadership in renewable energy was recognised by The Climate Group when it awarded former Governor Arnold Schwarzenegger its inaugural award for international climate leadership in Copenhagen in 2009. In Australia, the state of South Australia - under the leadership of former Premier Mike Rann - has led the way with wind power comprising 26% of its electricity generation by the end of 2011, edging out coal fired generation for the first time. South Australia also has had the highest take-up per capita of household solar panels in Australia following the Rann Government's introduction of solar feed-in laws and educative campaign involving the installation of solar photovoltaic installations on the roofs of prominent public buildings, including the parliament, museum, airport and Adelaide Showgrounds pavilion and schools. Rann, Australia's first climate change minister, passed legislation in 2006 setting targets for renewable energy and emissions cuts, the first legislation in Australia to do so.

Also, in the European Union there is a clear trend of promoting policies encouraging investments and financing for sustainable energy in terms of energy efficiency, innovation in energy exploitation and development of renewable resources, with increased consideration of environmental aspects and sustainability.

In October 2018, the American Council for an Energy-Efficient Economy (ACEEE) released its annual "State Energy Efficiency Scorecard." The scorecard concluded that states and electric utility companies are continuing to expand energy efficiency measures in order to meet clean energy goals. In 2017, the U.S. spent $6.6 billion in electricity efficiency programs. $1.3 billion was spent on natural gas efficiency. These programs resulted in 27.3 million megawatt hours (MWh) of electricity saved.

RENEWABLE ENERGY AND SUSTAINABLE DEVELOPMENT

Societies around the world are on the verge of a profound and urgently necessary transformation in the way they produce and use energy. This shift is moving the world away from the consumption of fossil fuels (which cause climate change and other environmental and social challenges) toward cleaner, renewable forms of energy. Millions of people around the world already use renewable energy to generate electricity, heat and cool buildings, cook and provide mobility. Renewable energy is market-ready and price competitive with conventional sources in many jurisdictions, and met about 19% of the world's final energy demand in 2014.

Around the world, communities, islands, and cities have found that making the transition to 100% renewable energy is largely a matter of political will and that the required technologies already are at hand. An increasing number of governments at all levels and on all continents is setting ambitious targets for renewable energy, with an ever-growing number of jurisdictions aiming for 100% renewables. Local governments, in particular, are pioneering this movement and are becoming incubators of regionally appropriate best practices and policies.

The rapid deployment of renewable energy has been driven mainly by a wide range of objectives (drivers), which include advancing economic development, improving energy security, enhancing energy access and mitigating climate change. Altogether, these drivers might be described as the pursuit of sustainable development, where economic prosperity is advanced around the world while negative impacts are minimized. While such presumed benefits are widely cited as key drivers in political and energy debates, specific, documented evidence of such benefits remains rather limited for reasons including a lack of adequate conceptual frameworks, methodological challenges, and limited access to relevant data.

APPROPRIATE TECHNOLOGY

Appropriate technology is a movement (and its manifestations) encompassing technological choice and application that is small-scale, decentralized, labor-intensive, energy-efficient, environmentally sound, and locally autonomous. It was originally articulated as intermediate technology by the economist Dr. Ernst Friedrich "Fritz" Schumacher in his work. Both Schumacher and many modern-day proponents of appropriate technology also emphasize the technology as people-centered.

Appropriate technology has been used to address issues in a wide range of fields. Well-known examples of appropriate technology applications include: bike- and hand-powered water pumps (and other self-powered equipment), the universal nut sheller, self-contained solar lamps and streetlights, and passive solar building designs. Today appropriate technology is often developed using open source principles, which have led to open-source appropriate technology (OSAT) and thus many of the plans of the technology can be freely found on the Internet. OSAT has been proposed as a new model of enabling innovation for sustainable development.

Appropriate technology is most commonly discussed in its relationship to economic development and as an alternative to technology transfer of more capital-intensive technology from industrialized nations to developing countries. However, appropriate technology movements can be found in both developing and developed countries. In developed countries, the appropriate technology movement grew out of the energy crisis of the 1970s and focuses mainly on environmental and sustainability issues. Today the idea is multifaceted; in some contexts, appropriate technology can be described as the simplest level of technology that can achieve the intended purpose, whereas in others, it can refer to engineering that takes adequate consideration of social and environmental ramifications. The facets are connected through robustness and sustainable living.

Energy Generation and Uses

The term soft energy technology was coined by Amory Lovins to describe "appropriate" renewable energy. "Appropriate" energy technologies are especially suitable for isolated and small scale energy needs. Electricity can be provided from:

- Photovoltaic (PV) solar panels, and (large) Concentrating solar power plants. PV solar panels made from low-cost photovoltaic cells or PV-cells which have first been concentrated by a Luminescent solar concentrator-panel are also a good option. Especially companies as Solfocus make appropriate technology CSP plants which can be made from waste plastics polluting the surroundings.

- Solar thermal collector.

- Wind power (home do-it yourself turbines and larger-scale).

- Micro hydro, and pico hydro.

- Human-powered handwheel generators.

- Other zero emission generation methods.

Some intermediate technologies include:

- Bioalcohols as bioethanol, biomethanol and biobutanol. The first two require minor modifications to allow them to be used in conventional gasoline engines. The third requires no modifications at all.

- Vegetable oils which can be used only in internal combustion (Diesel) engines. Biofuels are locally available in many developing countries and can be cheaper than fossil fuels.

- Anaerobic digestion power plants.

- Biogas is another potential source of energy, particularly where there is an abundant supply of waste organic matter. A generator (running on biofuels) can be run more efficiently if combined with batteries and an inverter; this adds significantly to capital cost but reduces running cost, and can potentially make this a much cheaper option than the solar, wind and micro-hydro options.

- Dry animal dung fuel can also be used.

- Biochar is another similar energy source which can be obtained through charring of certain types of organic material (e.g. hazelnut shells, bamboo, chicken manure) in a pyrolysis unit. A similar energy source is terra preta nova.

Finally, urine can also be used as a basis to generate hydrogen (which is an energy carrier). Using urine, hydrogen production is 332% more energy efficient than using water.

Electricity distribution could be improved so to make use of a more structured electricity line arrangement and universal AC power plugs and sockets (e.g. the CEE 7/7 plug). In addition, a universal system of electricity provisioning (e.g. universal voltage, frequency, ampere; e.g. 230 V with 50 Hz), as well as perhaps a better mains power system (e.g. through the use of special systems as perfected single-wire earth returns; e.g. Tunisia's MALT-system, which features low costs and easy placement).

Electricity storage (which is required for autonomous energy systems) can be provided through appropriate technology solutions as deep-cycle and car-batteries (intermediate technology), long duration flywheels, electrochemical capacitors, compressed air energy storage (CAES), liquid nitrogen and pumped hydro. Many solutions for the developing world are sold as a single package, containing a (micro) electricity generation power plant and energy storage. Such packages are called remote-area power supply.

LED Lamp with GU10 twist lock fitting, intended to replace halogen reflector lamps.

- White LEDs and a source of renewable energy (such as solar cells) are used by the Light Up the World Foundation to provide lighting to poor people in remote areas, and provide significant benefits compared to the kerosene lamps which they replace.

- Organic LEDs made by roll-to-roll production are another source of cheap light that will be commercially available at low cost by 2015.

- Compact fluorescent lamps (as well as regular fluorescent lamps and LED-lightbulbs) can also be used as appropriate technology. Although they are less environmentally friendly then LED-lights, they are cheaper and still feature relative high efficiency (compared to incandescent lamps).

- The Safe bottle lamp is a safer kerosene lamp designed in Sri Lanka. Lamps as these allow relative long, mobile, lighting. The safety comes from a secure screw-on metal lid, and two flat sides which prevent it from rolling if knocked over. An alternative to fuel or oil-based lanterns is the Uday lantern, developed by Philips as part of its Lighting Africa project (sponsored by the World Bank Group).

- The Faraday flashlight is a LED flashlight which operates on a capacitor. Recharging can be done by manual winching or by shaking, hereby avoiding the need of any supplementary electrical system.

- HID-lamps finally can be used for lighting operations where regular LED-lighting or other lamps will not suffice. Examples are car headlights. Due to their high efficiency, they are quite environmental, yet costly, and they still require polluting materials in their production process.

Transportation

Human powered-vehicles include the bicycle (and the future bamboo bicycle), which provides general-purpose transportation at lower costs compared to motorized vehicles, and many advantages over walking, and the whirlwind wheelchair, which provides mobility for disabled people who cannot afford the expensive wheelchairs used in developed countries. Animal powered vehicles/ transport may also be another appropriate technology. Certain zero-emissions vehicles may be considered appropriate transportation technology, including compressed air cars, liquid nitrogen and hydrogen-powered vehicles. Also, vehicles with internal combustion engines may be converted to hydrogen or oxyhydrogen combustion.

A man uses a bicycle to cargo goods in Ouagadougou, Burkina Faso.

Bicycles can also be applied to commercial transport of goods to and from remote areas. An example of this is Karaba, a free-trade coffee co-op in Rwanda, which uses 400 modified bicycles to carry hundreds of pounds of coffee beans for processing. Other projects for developing countries include the redesign of cycle rickshaws to convert them to electric power. However recent reports suggest that these rickshaws are not plying on the roads.

Determining a Sustainable Approach

Features such as low cost, low usage of fossil fuels and use of locally available resources can give some advantages in terms of sustainability. For that reason, these technologies are sometimes used and promoted by advocates of sustainability and alternative technology.

Besides using natural, locally available resources (e.g. wood or adobe), waste materials imported from cities using conventional (and inefficient) waste management may be gathered and re-used to build a sustainable living environment. Use of these cities' waste material allows the gathering of a huge amount of building material at a low cost. When obtained, the materials may be recycled over and over in the own city/community, using the cradle to cradle design method. Locations where waste can be found include landfills, junkyards, on water surfaces and anywhere around towns or near highways. Organic waste that can be reused to fertilise plants can be found in sewages. Also, town districts and other places (e.g. cemeteries) that are subject of undergoing renovation or removal can be used for gathering materials as stone, concrete, or potassium.

RENEWABLE RESOURCE

A renewable resource is a natural resource which will replenish to replace the portion depleted by usage and consumption, either through natural reproduction or other recurring processes in a finite amount of time in a human time scale. Renewable resources are a part of Earth's natural environment and the largest components of its ecosphere. A positive life cycle assessment is a key indicator of a resource's sustainability.

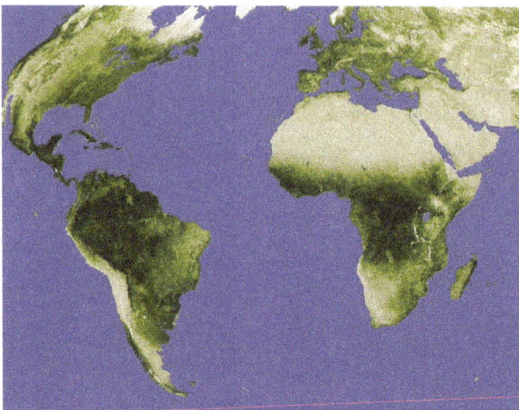

Global vegetation.

Oceans and seas often act as renewable resources.

Definitions of renewable resources may also include agricultural production, as in sustainable agriculture and to an extent water resources. In 1962, Paul Alfred Weiss defined renewable resources as: "The total range of living organisms providing man with life, fibres, etc". Another type of renewable resources is renewable energy resources. Common sources of renewable energy include solar, geothermal and wind power, which are all categorised as renewable resources.

Sawmill near Fügen, Zillertal, Austria.

Air, Food and Water

Water Resources

Water can be considered a *renewable* material when carefully controlled usage, treatment, and release are followed. If not, it would become a non-renewable resource at that location. For example, as groundwater is usually removed from an aquifer at a rate much greater than its very slow natural recharge, it is a considered non-renewable resource. Removal of water from the pore spaces in aquifers may cause permanent compaction (subsidence) that cannot be renewed. 97.5% of the water on the Earth is salt water, and 3% is fresh water; slightly over two thirds of this is frozen in glaciers and polar ice caps. The remaining unfrozen freshwater is found mainly as groundwater, with only a small fraction (0.008%) present above ground or in the air.

Water pollution is one of the main concerns regarding water resources. It is estimated that 22% of worldwide water is used in industry. Major industrial users include hydroelectric dams, thermoelectric power plants (which use water for cooling), ore and oil refineries (which use water in chemical processes) and manufacturing plants (which use water as a solvent).

Desalination of seawater is considered a renewable source of water, although reducing its dependence on fossil fuel energy is needed for it to be fully renewable.

Non Agricultural Food

Food is any substance consumed to provide nutritional support for the body. Most food has its origin in renewable resources. Food is obtained directly from plants and animals.

Alaska wild "berries" from the Innoko National Wildlife Refuge - Renewable Resources.

Hunting may not be the first source of meat in the modernised world, but it is still an important and essential source for many rural and remote groups. It is also the sole source of feeding for wild carnivores.

Sustainable Agriculture

The phrase sustainable agriculture was coined by Australian agricultural scientist Gordon Mc-Clymont. It has been defined as "an integrated system of plant and animal production practices having a site-specific application that will last over the long term". Expansion of agricultural land reduces biodiversity and contributes to deforestation. The Food and Agriculture Organization of the United Nations estimates that in coming decades, cropland will continue to be lost to industrial and urban development, along with reclamation of wetlands, and conversion of forest to cultivation, resulting in the loss of biodiversity and increased soil erosion.

Polyculture practices in Andhra Pradesh.

Although air and sunlight are available everywhere on Earth, crops also depend on soil nutrients and the availability of water. Monoculture is a method of growing only one crop at a time in a given field, which can damage land and cause it to become either unusable or suffer from reduced yields. Monoculture can also cause the build-up of pathogens and pests that target one specific species. The Great Irish Famine is a well-known example of the dangers of monoculture.

Crop rotation and long-term crop rotations confer the replenishment of nitrogen through the use of green manure in sequence with cereals and other crops, and can improve soil structure and fertility by alternating deep-rooted and shallow-rooted plants. Other methods to combat lost soil nutrients are returning to natural cycles that annually flood cultivated lands (returning lost nutrients indefinitely) such as the Flooding of the Nile, the long-term use of biochar, and use of crop and livestock landraces that are adapted to less than ideal conditions such as pests, drought, or lack of nutrients.

Agricultural practices are one of the single greatest contributor to the global increase in soil erosion rates. It is estimated that "more than a thousand million tonnes of southern Africa's soil are eroded every year. Experts predict that crop yields will be halved within thirty to fifty years if erosion continues at present rates." The Dust Bowl phenomenon in the 1930s was caused by severe drought combined with farming methods that did not include crop rotation, fallow fields, cover crops, soil terracing and wind-breaking trees to prevent wind erosion.

The tillage of agricultural lands is one of the primary contributing factors to erosion, due to mechanised agricultural equipment that allows for deep plowing, which severely increases the amount of soil that is available for transport by water erosion. The phenomenon called *peak soil* describes how large-scale factory farming techniques are affecting humanity's ability to grow food in the future. Without efforts to improve soil management practices, the availability of arable soil may become increasingly problematic.

Illegal slash and burn practice in Madagascar.

Methods to combat erosion include no-till farming, using a keyline design, growing wind breaks to hold the soil, and widespread use of compost. Fertilizers and pesticides can also have an effect of soil erosion,which can contribute to soil salinity and prevent other species from growing. Phosphate is a primary component in the chemical fertiliser applied most commonly in modern agricultural production. However, scientists estimate that rock phosphate reserves will be depleted in 50–100 years and that *Peak Phosphate* will occur in about 2030.

Industrial processing and logistics also have an effect on agriculture's sustainability. The way and locations crops are sold requires energy for transportation, as well as the energy cost for materials, labour, and transport. Food sold at a local location, such a farmers' market, have reduced energy overheads.

Air

Air is a renewable resource. All living organisms need oxygen, nitrogen (directly or indirectly), carbon (directly or indirectly) and many other gases in small quantities for their survival.

Examples of Industrial Use

Biorenewable Chemicals

Biorenewable chemicals are chemicals created by biological organisms that provide feedstocks for the chemical industry. Biorenewable chemicals can provide solar-energy-powered substitutes for the petroleum-based carbon feedstocks that currently supply the chemical industry. The tremendous diversity of enzymes in biological organisms, and the potential for synthetic biology to alter these enzymes to create yet new chemical functionalities, can drive the chemical industry. A major platform for creation of new chemicals is the polyketide biosynthetic pathway, which generates chemicals containing repeated alkyl chain units with potential for a wide variety of functional groups at the different carbon atoms.

Bioplastics

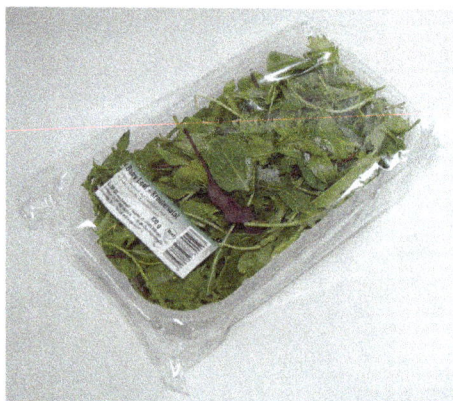

A packaging blister made from cellulose acetate, a bioplastic.

Bioplastics are a form of plastics derived from renewable biomass sources, such as vegetable fats and oils, lignin, corn starch, pea starch or microbiota. The most common form of bioplastic is thermoplastic starch. Other forms include Cellulose bioplastics, biopolyester, Polylactic acid, and bio-derived polyethylene.

The production and use of bioplastics is generally regarded as a more sustainable activity when compared to plastic production from petroleum (petroplastic); however, manufacturing of bioplastic materials is often still reliant upon petroleum as an energy and materials source. Because of the fragmentation in the market and ambiguous definitions it is difficult to describe the total market size for bioplastics, but the global production capacity is estimated at 327,000 tonnes. In contrast, global consumption of all flexible packaging is estimated at around 12.3 million tonnes.

Bioasphalt

Bioasphalt is an asphalt alternative made from non-petroleum based renewable resources. Manufacturing sources of bioasphalt include sugar, molasses and rice, corn and potato starches, and vegetable oil based waste. Asphalt made with vegetable oil based binders was patented by Colas SA in France in 2004.

Renewable Energy

Renewable energy refers to the provision of energy via renewable resources which are naturally replenished fast enough as being used. It includes e.g. sunlight, wind, biomass, rain, tides, waves and geothermal heat. Renewable energy may replace or enhance fossil energy supply various distinct areas: electricity generation, hot water/space heating, motor fuels, and rural (off-grid) energy services.

Biomass

Biomass is referring to biological material from living, or recently living organisms, most often referring to plants or plant-derived materials.

A sugarcane plantation in Brazil. Cane is used for biomass energy.

Sustainable harvesting and use of renewable resources (i.e., maintaining a positive renewal rate) can reduce air pollution, soil contamination, habitat destruction and land degradation. Biomass energy is derived from six distinct energy sources: garbage, wood, plants, waste, landfill gases, and alcohol fuels. Historically, humans have harnessed biomass-derived energy since the advent of burning wood to make fire, and wood remains the largest biomass energy source today.

However, low tech use of biomass, which still amounts for more than 10% of world energy needs may induce indoor air pollution in developing nations and results in between 1.5 million and 2 million deaths in 2000.

The biomass used for electricity generation varies by region. Forest by-products, such as wood residues, are common in the United States. Agricultural waste is common in Mauritius (sugar cane residue) and Southeast Asia (rice husks). Animal husbandry residues, such as poultry litter, are common in the UK. The biomass power generating industry in the United States, which consists of approximately 11,000 MW of summer operating capacity actively supplying power to the grid, produces about 1.4 percent of the U.S. electricity supply.

Biofuel

A biofuel is a type of fuel whose energy is derived from biological carbon fixation. Biofuels include fuels derived from biomass conversion, as well as solid biomass, liquid fuels and various biogases.

Bioethanol is an alcohol made by fermentation, mostly from carbohydrates produced in sugar or starch crops such as corn, sugarcane or switchgrass.

Brazil has bioethanol made from sugarcane available throughout the country. Shown a typical Petrobras gas station at São Paulo with dual fuel service, marked A for alcohol (ethanol) and G for gasoline.

Biodiesel is made from vegetable oils and animal fats. Biodiesel is produced from oils or fats using transesterification and is the most common biofuel in Europe. Biogas is methane produced by the process of anaerobic digestion of organic material by anaerobes., etc. is also a renewable source of energy.

Biogas

Biogas typically refers to a mixture of gases produced by the breakdown of organic matter in the absence of oxygen. Biogas is produced by anaerobic digestion with anaerobic bacteria or fermentation of biodegradable materials such as manure, sewage, municipal waste, green waste, plant material, and crops. It is primarily methane (CH_4) and carbon dioxide (CO_2) and may have small amounts of hydrogen sulphide (H_2S), moisture and siloxanes.

Natural Fibre

Natural fibres are a class of hair-like materials that are continuous filaments or are in discrete elongated pieces, similar to pieces of thread. They can be used as a component of composite materials. They can also be matted into sheets to make products such as paper or felt. Fibres are of two types: natural fibre which consists of animal and plant fibres, and man made fibre which consists of synthetic fibres and regenerated fibres.

Threats to Renewable Resources

Renewable resources are endangered by non-regulated industrial developments and growth. They must be carefully managed to avoid exceeding the natural world's capacity to replenish them. A life cycle assessment provides a systematic means of evaluating renewability. This is a matter of sustainability in the natural environment.

Overfishing

National Geographic has described ocean over fishing as "simply the taking of wildlife from the sea at rates too high for fished species to replace themselves."

Tuna meat is driving overfishing as to endanger some species like the bluefin tuna. The European Community and other organisations are trying to regulate fishery as to protect species and to prevent their extinctions. The United Nations Convention on the Law of the Sea treaty deals with aspects of overfishing in articles 61, 62, and 65.

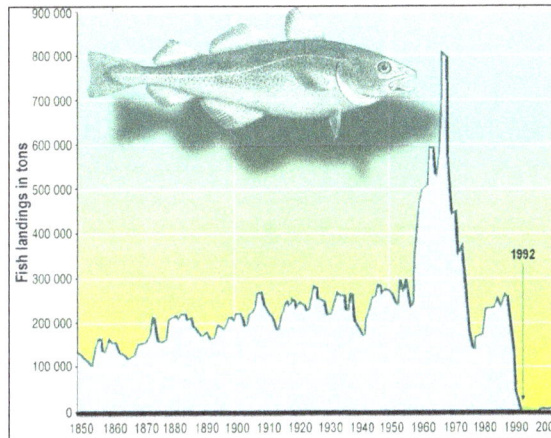

Atlantic cod stocks severely overfished leading to abrupt collapse.

Examples of overfishing exist in areas such as the North Sea of Europe, the Grand Banks of North America and the East China Sea of Asia. The decline of penguin population is caused in part by overfishing, caused by human competition over the same renewable resources.

Deforestation in Brazil 1996.

Deforestation

Besides their role as a resource for fuel and building material, trees protect the environment by absorbing carbon dioxide and by creating oxygen. The destruction of rain forests is one of the critical causes of climate change. Deforestation causes carbon dioxide to linger in the atmosphere. As carbon dioxide accrues, it produces a layer in the atmosphere that traps radiation from the sun. The radiation converts to heat which causes global warming, which is better known as the greenhouse effect.

Deforestation also affects the water cycle. It reduces the content of water in the soil and groundwater as well as atmospheric moisture. Deforestation reduces soil cohesion, so that erosion, flooding and landslides ensue.

Rain forests house many species and organisms providing people with food and other commodities. In this way biofuels may well be unsustainable if their production contributes to deforestation.

Endangered Species

Over-hunting of American Bison.

Some renewable resources, species and organisms are facing a very high risk of extinction caused by growing human population and over-consumption. It has been estimated that over 40% of all living species on Earth are at risk of going extinct. Many nations have laws to protect hunted species and to restrict the practice of hunting. Other conservation methods include restricting land development or creating preserves. The IUCN Red List of Threatened Species is the best-known worldwide conservation status listing and ranking system. Internationally, 199 countries have signed an accord agreeing to create Biodiversity Action Plans to protect endangered and other threatened species.

Renewable Energy

Renewable energy is energy that is collected from renewable resources, which are naturally replenished on a human timescale, such as sunlight, wind, rain, tides, waves, and geothermal heat. Renewable energy often provides energy in four important areas: electricity generation, air and water heating/cooling, transportation, and rural (off-grid) energy services.

Wind, solar, and hydroelectricity are three renewable sources of energy.

Based on REN21's 2017 report, renewables contributed 19.3% to humans' global energy consumption and 24.5% to their generation of electricity in 2015 and 2016, respectively. This energy consumption is divided as 8.9% coming from traditional biomass, 4.2% as heat energy (modern biomass, geothermal and solar heat), 3.9% from hydroelectricity and the remaining 2.2% is electricity from wind, solar, geothermal, and other forms of biomass. Worldwide investments in renewable technologies amounted to more than US$286 billion in 2015. In 2017, worldwide investments in renewable energy amounted to US$279.8 billion, with China accounting for US$126.6 billion or 45% of the global investments, the United States for US$40.5 billion, and Europe for US$40.9 billion. Globally, there are an estimated 7.7 million jobs associated with the renewable energy industries, with solar photovoltaics being the largest renewable employer. Renewable energy systems are rapidly becoming more efficient and cheaper and their share of total energy consumption is increasing. As of 2019 worldwide, more than two-thirds of all new electricity capacity installed was renewable. Growth in consumption of coal and oil could end by 2020 due to increased uptake of renewables and natural gas.

At the national level, at least 30 nations around the world already have renewable energy contributing more than 20 percent of energy supply. National renewable energy markets are projected to continue to grow strongly in the coming decade and beyond. Some places and at least two countries, Iceland and Norway, generate all their electricity using renewable energy already, and many other countries have the set a goal to reach 100% renewable energy in the future. At least 47 nations around the world already have over 50 percent of electricity from renewable resources. Renewable energy resources exist over wide geographical areas, in contrast to fossil fuels, which are concentrated in a limited number of countries. Rapid deployment of renewable energy and energy efficiency technologies is resulting in significant energy security, climate change mitigation, and economic benefits. In international public opinion surveys there is strong support for promoting renewable sources such as solar power and wind power.

While many renewable energy projects are large-scale, renewable technologies are also suited to rural and remote areas and developing countries, where energy is often crucial in human development. As most of renewable energy technologies provide electricity, renewable energy deployment is often applied in conjunction with further electrification, which has several benefits: electricity can be converted to heat (where necessary generating higher temperatures than fossil fuels), can be converted into mechanical energy with high efficiency, and is clean at the point of consumption. In addition, electrification with renewable energy is more efficient and therefore leads to significant reductions in primary energy requirements.

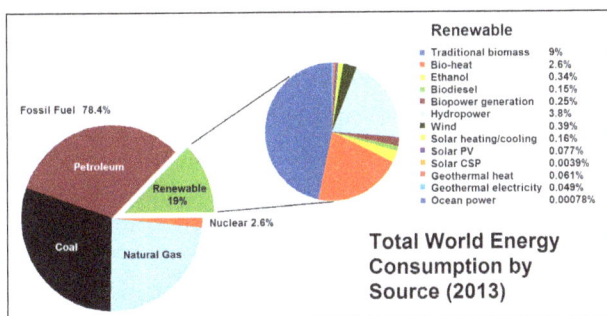

World energy consumption by source. Renewables accounted for 19% in 2012.

PlanetSolar, the world's largest solar-powered boat and the first ever solar electric vehicle to circumnavigate the globe.

Renewable energy flows involve natural phenomena such as sunlight, wind, tides, plant growth, and geothermal heat, as the International Energy Agency explains:

> Renewable energy is derived from natural processes that are replenished constantly. In its various forms, it derives directly from the sun, or from heat generated deep within the earth. Included in the definition is electricity and heat generated from solar, wind, ocean, hydropower, biomass, geothermal resources, and biofuels and hydrogen derived from renewable resources.

Renewable energy resources and significant opportunities for energy efficiency exist over wide geographical areas, in contrast to other energy sources, which are concentrated in a limited number of countries. Rapid deployment of renewable energy and energy efficiency, and technological diversification of energy sources, would result in significant energy security and economic benefits. It would also reduce environmental pollution such as air pollution caused by burning of fossil fuels and improve public health, reduce premature mortalities due to pollution and save associated health costs that amount to several hundred billion dollars annually only in the United States. Renewable energy sources, that derive their energy from the sun, either directly or indirectly, such as hydro and wind, are expected to be capable of supplying humanity energy for almost another 1 billion years, at which point the predicted increase in heat from the Sun is expected to make the surface of the earth too hot for liquid water to exist.

Climate change and global warming concerns, coupled with the continuing fall in the costs of some renewable energy equipment, such as wind turbines and solar panels, are driving increased use of renewables. New government spending, regulation and policies helped the industry weather the global financial crisis better than many other sectors. As of 2019, however, according to the International Renewable Energy Agency, renewables overall share in the energy mix (including power, heat and transport) needs to grow six times faster, in order to keep the rise in average global temperatures "well below" 2.0 °C (3.6 °F) during the present century, compared to pre-industrial levels.

As of 2011, small solar PV systems provide electricity to a few million households, and micro-hydro configured into mini-grids serves many more. Over 44 million households use biogas made in household-scale digesters for lighting and cooking, and more than 166 million households rely on a new generation of more-efficient biomass cookstoves. Secretary-General Ban Ki-moon has said that renewable energy has the ability to lift the poorest nations to new levels of prosperity. At the national level, at least 30 nations around the world already have renewable energy contributing more than 20% of energy supply. National renewable energy markets are projected to continue to grow strongly in the coming decade and beyond, and some 120 countries have various policy targets for longer-term shares of renewable energy, including a 20% target of all electricity generated for the European Union by 2020. Some countries have much higher long-term policy targets of up to 100% renewables. Outside Europe, a diverse group of 20 or more other countries target renewable energy shares in the 2020–2030 time frame that range from 10% to 50%.

Renewable energy often displaces conventional fuels in four areas: electricity generation, hot water/space heating, transportation, and rural (off-grid) energy services:

- Power generation: By 2040, renewable energy is projected to equal coal and natural gas electricity generation. Several jurisdictions, including Denmark, Germany, the state of South

Australia and some US states have achieved high integration of variable renewables. For example, in 2015 wind power met 42% of electricity demand in Denmark, 23.2% in Portugal and 15.5% in Uruguay. Interconnectors enable countries to balance electricity systems by allowing the import and export of renewable energy. Innovative hybrid systems have emerged between countries and regions.

- Heating: Solar water heating makes an important contribution to renewable heat in many countries, most notably in China, which now has 70% of the global total (180 GWth). Most of these systems are installed on multi-family apartment buildings and meet a portion of the hot water needs of an estimated 50–60 million households in China. Worldwide, total installed solar water heating systems meet a portion of the water heating needs of over 70 million households. The use of biomass for heating continues to grow as well. In Sweden, national use of biomass energy has surpassed that of oil. Direct geothermal for heating is also growing rapidly. The newest addition to Heating is from Geothermal Heat Pumps which provide both heating and cooling, and also flatten the electric demand curve and are thus an increasing national priority.

- Transportation: Bioethanol is an alcohol made by fermentation, mostly from carbohydrates produced in sugar or starch crops such as corn, sugarcane, or sweet sorghum. Cellulosic biomass, derived from non-food sources such as trees and grasses is also being developed as a feedstock for ethanol production. Ethanol can be used as a fuel for vehicles in its pure form, but it is usually used as a gasoline additive to increase octane and improve vehicle emissions. Bioethanol is widely used in the USA and in Brazil. Biodiesel can be used as a fuel for vehicles in its pure form, but it is usually used as a diesel additive to reduce levels of particulates, carbon monoxide, and hydrocarbons from diesel-powered vehicles. Biodiesel is produced from oils or fats using transesterification and is the most common biofuel in Europe.

A bus fueled by biodiesel.

A solar vehicle is an electric vehicle powered completely or significantly by direct solar energy. Usually, photovoltaic (PV) cells contained in solar panels convert the sun's energy directly into electric energy. The term "solar vehicle" usually implies that solar energy is used to power all or part of a vehicle's propulsion. Solar power may be also used to provide power for communications or controls or other auxiliary functions. Solar vehicles are not sold as practical day-to-day transportation devices at present, but are primarily demonstration vehicles and engineering exercises, often sponsored by government agencies. High-profile examples include PlanetSolar and Solar Impulse. However, indirectly solar-charged vehicles are widespread and solar boats are available commercially.

Prior to the development of coal in the mid 19th century, nearly all energy used was renewable. Almost without a doubt the oldest known use of renewable energy, in the form of traditional biomass to fuel fires, dates from more than a million years ago. Use of biomass for fire did not become commonplace until many hundreds of thousands of years later. Probably the second oldest usage of renewable energy is harnessing the wind in order to drive ships over water. This practice can be traced back some 7000 years, to ships in the Persian Gulf and on the Nile. From hot springs, geothermal energy has been used for bathing since Paleolithic times and for space heating since ancient Roman times. Moving into the time of recorded history, the primary sources of traditional renewable energy were human labor, animal power, water power, wind, in grain crushing windmills, and firewood, a traditional biomass.

In the 1860s and 1870s there were already fears that civilization would run out of fossil fuels and the need was felt for a better source. In 1873 Professor Augustin Mouchot wrote:

> The time will arrive when the industry of Europe will cease to find those natural resources, so necessary for it. Petroleum springs and coal mines are not inexhaustible but are rapidly diminishing in many places. Will man, then, return to the power of water and wind? Or will he emigrate where the most powerful source of heat sends its rays to all? History will show what will come.

In 1885, Werner von Siemens, commenting on the discovery of the photovoltaic effect in the solid state, wrote:

> We would say that however great the scientific importance of this discovery may be, its practical value will be no less obvious when we reflect that the supply of solar energy is both without limit and without cost, and that it will continue to pour down upon us for countless ages after all the coal deposits of the earth have been exhausted and forgotten.

Max Weber mentioned the end of fossil fuel in the concluding paragraphs of his Die protestantische Ethik und der Geist des Kapitalismus (The Protestant Ethic and the Spirit of Capitalism), published in 1905. Development of solar engines continued until the outbreak of World War I. The importance of solar energy was recognized in a 1911 article: "in the far distant future, natural fuels having been exhausted (solar power) will remain as the only means of existence of the human race".

The theory of peak oil was published in 1956. In the 1970s environmentalists promoted the development of renewable energy both as a replacement for the eventual depletion of oil, as well as for an escape from dependence on oil, and the first electricity-generating wind turbines appeared. Solar had long been used for heating and cooling, but solar panels were too costly to build solar farms until 1980.

Mainstream Technologies

Wind Power

In 2018, worldwide installed capacity of wind power was 564 GW.

Air flow can be used to run wind turbines. Modern utility-scale wind turbines range from around 600 kW to 9 MW of rated power. The power available from the wind is a function of the cube of the wind speed, so as wind speed increases, power output increases up to the maximum output for the particular turbine. Areas where winds are stronger and more constant, such as offshore and high-altitude sites, are preferred locations for wind farms. Typically, full load hours of wind turbines vary between 16 and 57 percent annually, but might be higher in particularly favorable offshore sites.

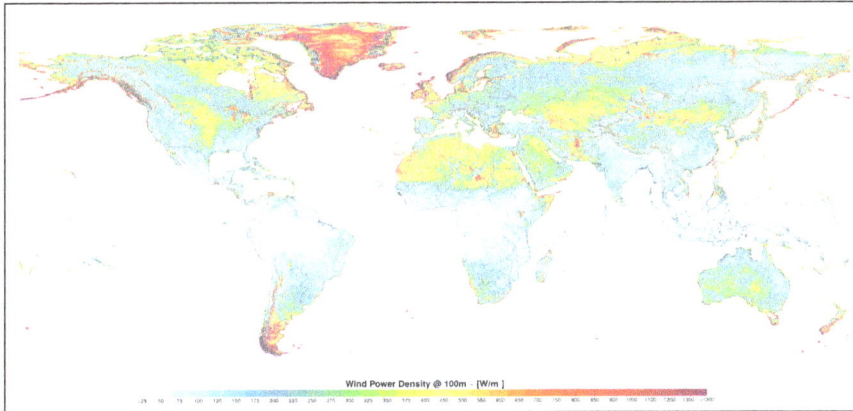

Global Map of Wind Power Density Potential.

Wind-generated electricity met nearly 4% of global electricity demand in 2015, with nearly 63 GW of new wind power capacity installed. Wind energy was the leading source of new capacity in Europe, the US and Canada, and the second largest in China. In Denmark, wind energy met more than 40% of its electricity demand while Ireland, Portugal and Spain each met nearly 20%.

Globally, the long-term technical potential of wind energy is believed to be five times total current global energy production, or 40 times current electricity demand, assuming all practical barriers needed were overcome. This would require wind turbines to be installed over large areas, particularly in areas of higher wind resources, such as offshore. As offshore wind speeds average ~90% greater than that of land, so offshore resources can contribute substantially more energy than land-stationed turbines.

Hydropower

In 2017, worldwide renewable hydropower capacity was 1,154 GW.

The Three Gorges Dam on the Yangtze River in China.

Since water is about 800 times denser than air, even a slow flowing stream of water, or moderate sea swell, can yield considerable amounts of energy. There are many forms of water energy:

- Historically, hydroelectric power came from constructing large hydroelectric dams and reservoirs, which are still popular in developing countries. The largest of them are the Three Gorges Dam in China and the Itaipu Dam built by Brazil and Paraguay.

- Small hydro systems are hydroelectric power installations that typically produce up to 50 MW of power. They are often used on small rivers or as a low-impact development on larger rivers. China is the largest producer of hydroelectricity in the world and has more than 45,000 small hydro installations.

- Run-of-the-river hydroelectricity plants derive energy from rivers without the creation of a large reservoir. The water is typically conveyed along the side of the river valley (using channels, pipes and tunnels) until it is high above the valley floor, whereupon it can allowed to fall through a penstock to drive a turbine. This style of generation may still produce a large amount of electricity, such as the Chief Joseph Dam on the Columbia river in the United States.

Hydropower is produced in 150 countries, with the Asia-Pacific region generating 32 percent of global hydropower in 2010. For countries having the largest percentage of electricity from renewables, the top 50 are primarily hydroelectric. China is the largest hydroelectricity producer, with 721 terawatt-hours of production in 2010, representing around 17 percent of domestic electricity use. There are now three hydroelectricity stations larger than 10 GW: the Three Gorges Dam in China, Itaipu Dam across the Brazil/Paraguay border, and Guri Dam in Venezuela.

Wave power, which captures the energy of ocean surface waves, and tidal power, converting the energy of tides, are two forms of hydropower with future potential; however, they are not yet widely employed commercially. A demonstration project operated by the Ocean Renewable Power Company on the coast of Maine, and connected to the grid, harnesses tidal power from the Bay of Fundy, location of world's highest tidal flow. Ocean thermal energy conversion, which uses the temperature difference between cooler deep and warmer surface waters, currently has no economic feasibility.

Geothermal Energy

Steam rising from the Nesjavellir Geothermal Power Station in Iceland.

Global geothermal capacity in 2017 was 12.9 GW.

High Temperature Geothermal energy is from thermal energy generated and stored in the Earth. Thermal energy is the energy that determines the temperature of matter. Earth's geothermal energy originates from the original formation of the planet and from radioactive decay of minerals (in currently uncertain but possibly roughly equal proportions). The geothermal gradient, which is the difference in temperature between the core of the planet and its surface, drives a continuous conduction of thermal energy in the form of heat from the core to the surface.

The heat that is used for geothermal energy can be from deep within the Earth, all the way down to Earth's core – 4,000 miles (6,400 km) down. At the core, temperatures may reach over 9,000 °F (5,000 °C). Heat conducts from the core to surrounding rock. Extremely high temperature and pressure cause some rock to melt, which is commonly known as magma. Magma convects upward since it is lighter than the solid rock. This magma then heats rock and water in the crust, sometimes up to 700 °F (371 °C).

Low Temperature Geothermal refers to the use of the outer crust of the earth as a Thermal Battery to facilitate Renewable thermal energy for heating and cooling buildings, and other refrigeration and industrial uses. In this form of Geothermal, a Geothermal Heat Pump and Ground-coupled heat exchanger are used together to move heat energy into the earth (for cooling) and out of the earth (for heating) on a varying seasonal basis. Low temperature Geothermal (generally referred to as "GHP") is an increasingly important renewable technology because it both reduces total annual energy loads associated with heating and cooling, and it also flattens the electric demand curve eliminating the extreme summer and winter peak electric supply requirements. Thus Low Temperature Geothermal/GHP is becoming an increasing national priority with multiple tax credit support and focus as part of the ongoing movement toward Net Zero Energy.

Bioenergy

Biomass is biological material derived from living, or recently living organisms. It most often refers to plants or plant-derived materials which are specifically called lignocellulosic biomass. As an energy source, biomass can either be used directly via combustion to produce heat, or indirectly after converting it to various forms of biofuel. Conversion of biomass to biofuel can be achieved by different methods which are broadly classified into: thermal, chemical, and biochemical methods. Wood remains the largest biomass energy source today; examples include forest residues – such as dead trees, branches and tree stumps – yard clippings, wood chips and even municipal solid waste. In the second sense, biomass includes plant or animal matter that can be converted into fibers or other industrial chemicals, including biofuels. Industrial biomass can be grown from numerous types of plants, including miscanthus, switchgrass, hemp, corn, poplar, willow, sorghum, sugarcane, bamboo, and a variety of tree species, ranging from eucalyptus to oil palm (palm oil).

Plant energy is produced by crops specifically grown for use as fuel that offer high biomass output per hectare with low input energy. The grain can be used for liquid transportation fuels while the straw can be burned to produce heat or electricity. Plant biomass can also be degraded from cellulose to glucose through a series of chemical treatments, and the resulting sugar can then be used as a first generation biofuel.

Biomass can be converted to other usable forms of energy such as methane gas or transportation fuels such as ethanol and biodiesel. Rotting garbage, and agricultural and human waste, all release methane gas – also called landfill gas or biogas. Crops, such as corn and sugarcane, can be fermented to produce the transportation fuel, ethanol. Biodiesel, another transportation fuel, can be produced from left-over food products such as vegetable oils and animal fats. Also, biomass to liquids (BTLs) and cellulosic ethanol are still under research. There is a great deal of research involving algal fuel or algae-derived biomass due to the fact that it is a non-food resource and can be produced at rates 5 to 10 times those of other types of land-based agriculture, such as corn and soy. Once harvested, it can be fermented to produce biofuels such as ethanol, butanol, and methane, as well as biodiesel and hydrogen. The biomass used for electricity generation varies by region. Forest by-products, such as wood residues, are common in the United States. Agricultural waste is common in Mauritius (sugar cane residue) and Southeast Asia (rice husks). Animal husbandry residues, such as poultry litter, are common in the United Kingdom.

Biofuels include a wide range of fuels which are derived from biomass. The term covers solid, liquid, and gaseous fuels. Liquid biofuels include bioalcohols, such as bioethanol, and oils, such as biodiesel. Gaseous biofuels include biogas, landfill gas and synthetic gas. Bioethanol is an alcohol made by fermenting the sugar components of plant materials and it is made mostly from sugar and starch crops. These include maize, sugarcane and, more recently, sweet sorghum. The latter crop is particularly suitable for growing in dryland conditions, and is being investigated by International Crops Research Institute for the Semi-Arid Tropics for its potential to provide fuel, along with food and animal feed, in arid parts of Asia and Africa.

With advanced technology being developed, cellulosic biomass, such as trees and grasses, are also used as feedstocks for ethanol production. Ethanol can be used as a fuel for vehicles in its pure form, but it is usually used as a gasoline additive to increase octane and improve vehicle emissions. Bioethanol is widely used in the United States and in Brazil. The energy costs for producing bio-ethanol are almost equal to, the energy yields from bio-ethanol. However, according to the European Environment Agency, biofuels do not address global warming concerns. Biodiesel is made from vegetable oils, animal fats or recycled greases. It can be used as a fuel for vehicles in its pure form, or more commonly as a diesel additive to reduce levels of particulates, carbon monoxide, and hydrocarbons from diesel-powered vehicles. Biodiesel is produced from oils or fats using transesterification and is the most common biofuel in Europe. Biofuels provided 2.7% of the world's transport fuel in 2010.

Biomass, biogas and biofuels are burned to produce heat/power and in doing so harm the environment. Pollutants such as sulphurous oxides (SO_x), nitrous oxides (NO_x), and particulate matter (PM) are produced from the combustion of biomass; the World Health Organisation estimates that 7 million premature deaths are caused each year by air pollution. Biomass combustion is a major contributor.

Integration into the Energy System

Renewable energy production from some sources such as wind and solar is more variable and more geographically spread than technology based on fossil fuels and nuclear. While integrating it into the wider energy system is feasible, it does lead to some additional challenges. In order for the energy system to remain stable, a set of measurements can be taken. Implementation of energy storage, using a wide variety of renewable energy technologies, and implementing a smart grid in which energy is automatically used at the moment it is produced can reduce risks and costs of renewable energy implementation.

Electrical Energy Storage

Electrical energy storage is a collection of methods used to store electrical energy. Electrical energy is stored during times when production (especially from intermittent sources such as wind power, tidal power, solar power) exceeds consumption, and returned to the grid when production falls below consumption. Pumped-storage hydroelectricity accounts for more than 90% of all grid power storage. Costs of lithium-ion batteries are dropping rapidly, and are increasingly being deployed grid ancillary services and for domestic storage.

CLEAN TECHNOLOGY

Clean technology is any process, product, or service that reduces negative environmental impacts through significant energy efficiency improvements, the sustainable use of resources, or environmental protection activities. Clean technology includes a broad range of technology related to recycling, renewable energy (wind energy, solar energy, biomass, hydropower, biofuels, etc.), information technology, green transportation, electric motors, green chemistry, lighting, Greywater, and more. Environmental finance is a method by which new clean technology projects that have proven that they are "additional" or "beyond business as usual" can obtain financing through the generation of carbon credits. A project that is developed with concern for climate change mitigation (such as a Kyoto Clean Development Mechanism project) is also known as a carbon project.

Clean Edge, a clean technology research firm, describes clean technology "a diverse range of products, services, and processes that harness renewable materials and energy sources, dramatically reduce the use of natural resources, and cut or eliminate emissions and wastes." Clean Edge notes that, "Clean technologies are competitive with, if not superior to, their conventional counterparts. Many also offer significant additional benefits, notably their ability to improve the lives of those in both developed and developing countries".

Investments in clean technology have grown considerably since coming into the spotlight around 2000. According to the United Nations Environment Program, wind, solar, and biofuel companies received a record $148 billion in new funding in 2007 as rising oil prices and climate change policies encouraged investment in renewable energy. $50 billion of that funding went to wind power. Overall, investment in clean-energy and energy-efficiency industries rose 60 percent from 2006 to 2007. In 2009, it was forecast that the three main clean technology sectors, solar photovoltaics, wind power, and biofuels, will have revenues of $325.1 billion by 2018.

Cleantech products or services are those that improve operational performance, productivity, or efficiency while reducing costs, inputs, energy consumption, waste, or environmental pollution. Its origin is the increased consumer, regulatory, and industry interest in clean forms of energy generation—specifically, perhaps, the rise in awareness of global warming, climate change, and the impact on the natural environment from the burning of fossil fuels. Cleantech is often associated with venture capital funds and land use organizations. The term has historically been differentiated from various definitions of green business, sustainability, or triple bottom line industries by its origins in the venture capital investment community and has grown to define a business sector that includes significant and high growth industries such as solar, wind, water purification, and biofuels.

PURPOSE OF CLEAN ENERGY

Clean energy development is vital for combating climate change and limiting its most devastating effects. 2014 was the warmest year on record. The Earth's temperature has risen by an average 0.85 °C since the end of the 19th Century, states National Geographic in its special November 2015 issue on climate change.

Meanwhile, some 1.1 billion inhabitants (17% of the world population) do not have access to electricity. Equally, 2.7 billion people (38% of the population) use conventional biomass for cooking, heating and lighting in their homes - at serious risk to their health.

As such, one of the objectives established by the United Nations is to achieve to access to electricity for everyone by 2030, an ambitious target considering that, by then, according to the IEA's estimates, 800 million people will have no access to an electricity supply if current trends continue.

Renewable energies received important backing from the international community through the Paris Accord signed at the World Climate Summit held in the French capital in December 2015.

The agreement, which will enter into force in 2020, establishes, for the first time in history, a binding global objective. Nearly 200 signatory countries pledged to reduce their emissions so that the average temperature of the planet at the end of the current century remains "well below" 2 °C, the limit above which climate change will have more catastrophic effects. The aim is to try to keep it to 1.5 °C.

Likewise, the transition to an energy system based on renewable technologies will have very positive economic consequences. doubling the renewable energy share in the world energy mix, to 36% by 2030, will result in additional global growth of 1.1% by that year (equivalent to 1.3 trillion dollars), a increase in wellbeing of 3.7% and in employment in the sector of up to more than 24 million people, compared to 9.2 million today.

References

- Sustainable-technologies-matter: brightsync.green, Retrieved 3 May, 2019

- Ricardo David Lopes (1 July 2010). "Primeiro parque mundial de ondas na Póvoa de Varzim". Jn.sapo.pt. Retrieved 8 July2010

- Green-technology: brightsync.green, Retrieved 4 April, 2019

- Pennwell Corporation (2008-07-08). "Uday lamp and lighting africa project description". Ledsmagazine.com. Retrieved 2012-07-28

- Sustainableenergy: conserve-energy-future.com, Retrieved 5 May, 2019

- Cordell; et al. (2009-02-11). "The story of phosphorus: Global food security and food for thought". Global Environmental Change. 19 (2): 292–305. Doi:10.1016/j.gloenvcha.2008.10.009

- Renewable-energy: acciona.com, Retrieved 6 June 2019

- Spellman, Frank R. (2013). Safe Work Practices for Green Energy Jobs (first ed.). Destech Publications. p. 323. ISBN 978-1-60595-075-4. Retrieved 29 December 2014

Sustainable Waste Management

Sustainable waste management reduces the amount of natural resources consumed and ensures that any materials that are taken from nature are reused as many times as possible. This chapter has been carefully written to provide an easy understanding of the varied facets of sustainable waste management.

Solid-waste management is the collection treating, and disposing of solid material that is discarded because it has served its purpose or is no longer useful. Improper disposal of municipal solid waste can create unsanitary conditions, and these conditions in turn can lead to pollution of the environment and to outbreaks of vector-borne disease—that is, diseases spread by rodents and insects. The tasks of solid-waste management present complex technical challenges. They also pose a wide variety of administrative, economic, and social problems that must be managed and solved.

Sustainable Practices in Waste Management

Ways to Create an Efficient Waste Management Plan

You can create an efficient plan for waste management in your facility through the following 4 ways:

Considering Sustainable Materials Management

Don't consider waste management as your last resort to manage waste efficiently; rather take the approach of sustainable materials management. The former needs you to look at all the waste that is generated and think of different methods in which you can recycle or reuse the waste. However, the latter allows you to make deliberate and informed decisions about how materials should flow at different manufacturing stages to generate less waste.

Planning at Every Stage

Planning for waste management is not a one-time event but a process consisting of various stages that come together to help you achieve your goals. Follow and track your plan at every stage. By employing strategic planning, you get the opportunity to deliver sustainable improvements to local waste management practices as it has the ability to respond to the ever changing waste and recovered materials markets.

Collaborating Whenever Possible

You must collaborate with different organizations and companies that share the same goal. Public-Private Partnerships for Service Delivery (PPPSD) is one such approach that promotes sustainable and self-supporting partnerships between various businesses and local governments.

This kind of collaboration helps in stimulating an improved cooperation between public, private and citizen stakeholders. It also helps in minimizing the adverse effects of waste in poor communities, contributes to the sustainable improvement of recycling and solid waste management, and improves the livelihood of people and businesses in rural and urban communities equally.

Aiming to Avoid the Landfills

Aim to stray away from landfills as much as possible. Civic bodies must make an effort to operate under various legislative requirements that want to achieve specific diversion goals. Determine the actual diversion rate at the different stages of recycling programs. You must know the quantity of materials that was usable in the production of recyclable products.

Importance of Accurate Weighing in Recycling

Everyone in the recycling industry stresses the importance of accurate weighing of materials. In fact, we can say that the recycling industry depends on weighing recycling waste accurately, regardless of whether you are a buyer or a seller.

By incorporating weigh scales such as truck scales, forklift scales, floor scales, bench scales, etc., you can ensure that every waste material, no matter what it is made up of, is weighed accurately so that you know exactly how much is being recycled, reduced and sent to the landfills. It also helps in getting the right amount of money corresponding to the exact quantity you are selling or buying.

Different Stages of Recycling

There are 4 stages in every recycling facility. Let us take a brief look at what happens at these four stages:

Transfer Truck

At every collection station, all the recycling collection trucks discard the recyclable waste into a large transfer truck.

Loading into Hopper

As soon as the collection trucks arrive at the materials recovery facility, they unload the cans, bottles, cardboard, paper, and other recyclables. All these are put into a hopper using a front-end loader, which further feeds them to a conveyor belt.

Sorting

Now comes one of the most crucial tasks — sorting different recyclables into different piles.

The glass sorting rollers consist of slotted rollers that pull out and crush glass bottles. This glass can be used as an aggregate for developmental projects or sent to recycling facilities which use this glass to produce different objects.

The container/paper sorting machine helps in separating paper and cardboard. The conveyor feeds the remaining stream of recyclables (minus the glass) to a rake-like separator. This causes the plastic and metal containers to drop off to a side conveyor, while paper and cardboard roll over the top.

The fiber sorting lines are three conveyor lines where fiber items that include paper, boxboard and cardboard are pulled out by hand. They are then dropped into chutes over separate bins from where they are ready to be fed to a conveyor that goes to the fiber baler.

Different types of plastic bottles are pulled off by hand in the container sorting lines. They are then dropped into bins that go to the container baler

Finally, the magnetic can sorters remove everything that is steel/metal. Once the plastic and paper are removed from the container line, a rotating magnet above the line grabs all the steel cans and pulls then into a bin. Once the steel cans are removed, an eddy current separator is used to remove aluminum cans from the line. This separator uses a magnetized rotor under the conveyor which repels non-ferrous aluminum cans. This causes them to 'jump off' the belt and drop into a bin below.

Balers

The cardboard and paper that is separated in the previous stages are now fed to a fiber baler. It runs continuously where bales of paper and cardboard are stacked and stored. These bales are sold or shipped to paper mills where the recycled fiber is used to manufacture a number of different products such as newspapers, cardboard boxes, office papers, notebooks, etc.

The plastic containers and metal cans are fed to a separate compactor and baler for containers. Bales of compacted bi-metal cans, such as cans made up of steel and tin, and aluminum cans are stored and sold to metal processors.

These metals are then used to make a wide range of metal products. Bales of no. 1 plastics (PET), no. 2 plastics (HDPE), and mixed plastics are sold to manufacturers that make thousands of products including lumber, bottles, and carpeting.

Once all the recyclable materials are sorted, the remaining waste that cannot be recycled or reused is sent to the landfills.

SUSTAINABLE SOLID WASTE MANAGEMENT

The primary aim of sustainable solid waste management is to address concerns related to public health, environmental pollution, land use, resource management and socio-economic impacts associated with improper disposal of waste. "This growing mountain of garbage and trash represents not only an attitude of indifference toward valuable natural resources, but also a serious economic and public health problem". These words are is enough to understand the social, economical and environmental impact of mismanaged waste disposal and an urgent call for help to look for innovative, smart, sustainable and effective waste disposal techniques.

According to UNEP, around 3 billion tons of waste is generated every year, with industrial waste being the largest contributor, especially from China, EU and USA. There has been a steady increase in the quantity of e-wastes and hazardous waste materials. The UNEP study observed a drastic shift from high organic to higher plastic and paper corresponding to increase in the standards of living and also made an interesting correlation between the higher GDP and the quantity of municipal waste collections.

In developing and under-developed countries, the use of open dumps to dispose of the solid waste from different sectors is staggeringly high compared to the developed and high income countries that are more dependent on recycling and use of sanitary landfills that are isolated from the surrounding environment until it is safe.

There are serious concerns on the increasing cost of waste disposal, especially in developing countries. It is estimated that around $200 billion are being spent on waste management in the OECD countries for both municipal and industrial waste.

For developing countries, at least 20-50% of its annual budget is devoted to waste managementschemes and strategy that has been reported insufficient and inefficient at the same time. In these countries, use of unscientific and at times unethical and outdated waste management practices have led to various environmental repercussions and economic backlashes. Even the relatively small proportion of waste recycling and other waste minimization and re-use techniques for waste disposal is alarming.

The increasing cost of waste disposal is a cause of major concern in developing nations.

As sustainable solid waste management evolves through waste awareness among general public, efforts within the industry, and waste management becoming not just an environmental concern but a political and strategic apprehension too, there are realistic chances of advancements and scientific innovations.

Innovation will then give birth to revolutionary and self-sustaining ideas within the industry, which earlier focused on basic waste management, will now grow towards maximum utilization and sustainable management of waste.

In the last couple of decades, sustainable solid waste management has become a matter of political significance with robust policies, strategies and agendas devised to address the issue. The good thing is that the industry has responded with innovative, cost-effective and customized solutions to manage solid wastes in an environmental-friendly manner.

WASTE MINIMISATION

Waste minimisation is a set of processes and practices intended to reduce the amount of waste produced. By reducing or eliminating the generation of harmful and persistent wastes, waste minimisation supports efforts to promote a more sustainable society. Waste minimisation involves redesigning products and processes and changing societal patterns of consumption and production.

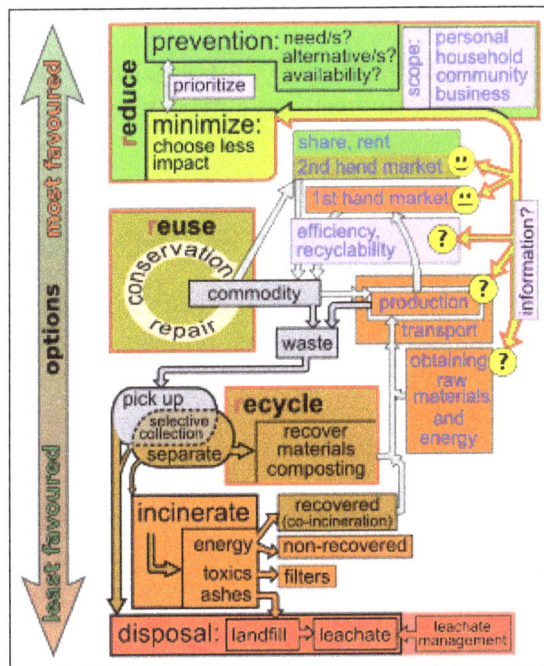

Waste hierarchy: refusing, reducing, reusing, recycling and composting allow to reduce waste.

The most environmentally resourceful, economically efficient, and cost effective way to manage waste often is to not have to address the problem in the first place. Managers see waste minimisation as a primary focus for most waste management strategies. Proper waste treatment and disposal can require a significant amount of time and resources; therefore, the benefits of waste minimisation can be considerable if carried out in an effective, safe and sustainable manner.

Traditional waste management focuses on processing waste after it is created, concentrating on reuse, recycling, and waste-to-energy conversion. Waste minimisation involves efforts to avoid creating the waste during manufacturing. To effectively implement waste minimisation the manager requires knowledge of the production process, cradle-to-grave analysis (the tracking of materials from their extraction to their return to earth) and details of the composition of the waste.

The main sources of waste vary from country to country. In the UK, most waste comes from the construction and demolition of buildings, followed by mining and quarrying, industry and commerce. Household waste constitutes a relatively small proportion of all waste. Industrial waste is often tied to requirements in the supply chain. For example, a company handling a product may insist that it should be shipped using particular packing because it fits downstream needs.

Benefits

Waste minimisation can protect the environment and often turns out to have positive economic benefits. Waste minimisation can improve:

- Efficient production practices: Waste minimisation can achieve more output of product per unit of input of raw materials.

- Economic returns: More efficient use of products means reduced costs of purchasing new materials improving the financial performance of a company.

- Public image: The environmental profile of a company is an important part of its overall reputation and waste minimisation reflects a proactive movement towards environmental protection.

- Quality of products produced: New innovation and technological practices can reduce waste generation and improve the quality of the inputs in the production phase.

- Environmental responsibility: Minimising or eliminating waste generation makes it easier to meet targets of environmental regulations, policies, and standards. The environmental impact of waste will be reduced.

Industries

In industry, using more efficient manufacturing processes and better materials generally reduces the production of waste. The application of waste minimisation techniques has led to the development of innovative and commercially successful replacement products.

Waste minimisation efforts often require investment, which is usually compensated by the savings. However, waste reduction in one part of the production process may create waste production in another part.

Processes

- Reuse of scrap material: Scraps can be immediately re-incorporated at the beginning of the manufacturing line so that they do not become a waste product. Many industries routinely do this; for example, paper mills return any damaged rolls to the beginning of the production line, and in the manufacture of plastic items, off-cuts and scrap are re-incorporated into new products.

- Improved quality control and process monitoring: Steps can be taken to ensure that the number of reject batches is kept to a minimum. This is achieved by increasing the frequency of inspection and the number of points of inspection. For example, installing automated continuous monitoring equipment can help to identify production problems at an early stage.

- Waste exchanges: This is where the waste product of one process becomes the raw material for a second process. Waste exchanges represent another way of reducing waste disposal volumes for waste that cannot be eliminated.

- Ship to point of use: This involves making deliveries of incoming raw materials or components direct to the point where they are assembled or used in the manufacturing process to minimise handling and the use of protective wrappings or enclosures.

- Zero waste: This is a whole systems approach that aims to eliminate waste at the source and at all points down the supply chain, with the intention of producing no waste. It is a design philosophy which emphasizes waste prevention as opposed to end of pipe waste management. Since, globally speaking, waste as such, however minimal, can never be prevented (there will always be an end-of-life even for recycled products and materials), a related goal is pollution prevention.

Product Design

Waste minimisation and resource maximisation for manufactured products can most easily be done at the design stage. Reducing the number of components used in a product or making the product easier to take apart can make it easier to be repaired or recycled at the end of its useful life.

In some cases, it may be best not to minimise the volume of raw materials used to make a product, but instead reduce the volume or toxicity of the waste created at the end of a product's life, or the environmental impact of the product's use.

Fitting the Intended Use

In this strategy, products and packages are optimally designed to meet their intended use. This applies especially to packaging materials, which should only be as durable as necessary to serve their intended purpose. On the other hand, it could be more wasteful if food, which has consumed resources and energy in its production, is damaged and spoiled because of extreme measures to reduce the use of paper, metals, glass and plastics in its packaging.

Durability

Improving product durability, such as extending a vacuum cleaner's useful life to 15 years instead of 12, can reduce waste and usually much improves resource optimisation.

But in some cases it has a negative environmental impact. If a product is too durable, its replacement with more efficient technology is likely to be delayed. Therefore, extending an older machine's useful life may place a heavier burden on the environment than scrapping it, recycling its metal and buying a new model. Similarly, older vehicles consume more fuel and produce more emissions than their modern counterparts.

Most proponents of waste minimisation consider that the way forward may be to view any manufactured product at the end of its useful life as a resource for recycling and reuse rather than waste.

Making refillable glass bottles strong enough to withstand several journeys between the consumer and the bottling plant requires making them thicker and so heavier, which increases the resources required to transport them. Since transport has a large environmental impact, careful evaluation is required of the number of return journeys bottles make. If a refillable bottle is thrown away after being refilled only several times, the resources wasted may be greater than if the bottle had been designed for a single journey.

Many choices involve trade-offs of environmental impact, and often there is insufficient information to make informed decisions.

Retail

Various aspects of business practices affect waste, such as the use of disposable tableware in restaurants.

Reusable Shopping Bags

Reusable bags are a visible form of re-use, and some stores offer a "bag credit" for re-usable shopping bags, although at least one chain reversed its policy, claiming "it was just a temporary bonus". In contrast, one study suggests that a bag tax is a more effective incentive than a similar discount. While there is a minor inconvenience involved, this may remedy itself, as reusable bags are generally more convenient for carrying groceries.

Households

Appropriate amounts and sizes can be chosen when purchasing goods; buying large containers of paint for a small decorating job or buying larger amounts of food than can be consumed create unnecessary waste. Also, if a pack or can is to be thrown away, any remaining contents must be removed before the container can be recycled.

Home composting, the practice of turning kitchen and garden waste into compost can be considered waste minimisation.

The resources that households use can be reduced considerably by using electricity thoughtfully (e.g. turning off lights and equipment when it is not needed) and by reducing the number of car journeys made. Individuals can reduce the amount of waste they create by buying fewer products and by buying products which last longer. Mending broken or worn items of clothing or equipment also contributes to minimising household waste. Individuals can minimise their water usage, and walk or cycle to their destination rather than using their car to save fuel and cut down emissions.

In a domestic situation, the potential for minimisation is often dictated by lifestyle. Some people may view it as wasteful to purchase new products solely to follow fashion trends when the older products are still usable. Adults working full-time have little free time, and so may have to purchase more convenient foods that require little preparation, or prefer disposable nappies if there is a baby in the family.

The amount of waste an individual produces is a small portion of all waste produced by society, and personal waste reduction can only make a small impact on overall waste volumes. Yet, influence on policy can be exerted in other areas. Increased consumer awareness of the impact and power of certain purchasing decisions allows industry and individuals to change the total resource consumption. Consumers can influence manufacturers and distributors by avoiding buying products that do not have eco-labelling, which is currently not mandatory, or choosing products that minimise the use of packaging. In the UK, PullApart combines both environmental and consumer packaging surveys, in a curbside packaging recycling classification system to minimise waste. Where reuse schemes are available, consumers can be proactive and use them.

Health-care Facilities

Health-care establishments are massive producers of waste. The major sources of health-care waste are: hospitals, laboratories and research centres, mortuary and autopsy centres, animal research and testing laboratories, blood banks and collection services, and nursing homes for the elderly.

Waste minimisation can offer many opportunities to these establishments to use fewer resources, be less wasteful and generate less hazardous waste. Good management and control practices among health-care facilities can have a significant effect on the reduction of waste generated each day.

Practices

There are many examples of more efficient practices that can encourage waste minimization in healthcare establishments and research facilities.

Source Reduction

- Purchasing reductions which ensures the selection of supplies that are less wasteful or less hazardous.

- The use of physical rather than chemical cleaning methods such as steam disinfection instead of chemical disinfection.

- Preventing the unnecessary wastage of products in nursing and cleaning activities.

Management and Control Measures at Hospital Level

- Centralized purchasing of hazardous chemicals.

- Monitoring the flow of chemicals within the health care facility from receipt as a raw material to disposal as a hazardous waste.

- The careful separation of waste matter to help minimize the quantities of hazardous waste and disposal.

Stock Management of Chemical and Pharmaceutical Products

- Frequent ordering of relatively small quantities rather than large quantities at one time.

- Using the oldest batch of a product first to avoid expiration dates and unnecessary waste.

- Using all the contents of a container containing hazardous waste.

- Checking the expiry date of all products at the time of delivery.

SEWAGE TREATMENT

Sewage treatment is the process of removing contaminants from municipal wastewater, containing mainly household sewage plus some industrial wastewater. Physical, chemical, and biological processes are used to remove contaminants and produce treated wastewater (or treated effluent) that is safe enough for release into the environment. A by-product of sewage treatment is a semi-solid waste or slurry, called sewage sludge. The sludge has to undergo further treatment before being suitable for disposal or application to land.

Sewage treatment plant in Massachusetts, US.

Sewage treatment may also be referred to as wastewater treatment. However, the latter is a broader term which can also refer to industrial wastewater. For most cities, the sewer system will also carry a proportion of industrial effluent to the sewage treatment plant which has usually received pre-treatment at the factories themselves to reduce the pollutant load. If the sewer system is a combined sewer then it will also carry urban runoff (stormwater) to the sewage treatment plant. Sewage water can travel towards treatment plants via piping and in a flow aided by gravity and pumps. The first part of filtration of sewage typically includes a bar screen to filter solids and large objects which are then collected in dumpsters and disposed of in landfills. Fat and grease is also removed before the primary treatment of sewage.

Process Steps

Sewage collection and treatment in the United States is typically subject to local, state and federal regulations and standards.

Treating wastewater has the aim to produce an effluent that will do as little harm as possible when discharged to the surrounding environment, thereby preventing pollution compared to releasing untreated wastewater into the environment.

Sewage treatment generally involves three stages, called primary, secondary and tertiary treatment.

- Primary treatment consists of temporarily holding the sewage in a quiescent basin where heavy solids can settle to the bottom while oil, grease and lighter solids float to the surface. The settled and floating materials are removed and the remaining liquid may be discharged or subjected to secondary treatment. Some sewage treatment plants that are connected to a combined sewer system have a bypass arrangement after the primary treatment unit. This means that during very heavy rainfall events, the secondary and tertiary treatment systems can be bypassed to protect them from hydraulic overloading, and the mixture of sewage and stormwater only receives primary treatment.

- Secondary treatment removes dissolved and suspended biological matter. Secondary treatment is typically performed by indigenous, water-borne micro-organisms in a managed habitat. Secondary treatment may require a separation process to remove the micro-organisms from the treated water prior to discharge or tertiary treatment.

- Tertiary treatment is sometimes defined as anything more than primary and secondary treatment in order to allow ejection into a highly sensitive or fragile ecosystem (estuaries, low-flow rivers, coral reefs, etc.). Treated water is sometimes disinfected chemically or physically (for example by lagoons and microfiltration) prior to discharge into a stream, river, bay, lagoon or wetland, or it can be used for the irrigation of a golf course, green way or park. If it is sufficiently clean, it can also be used for groundwater recharge or agricultural purposes.

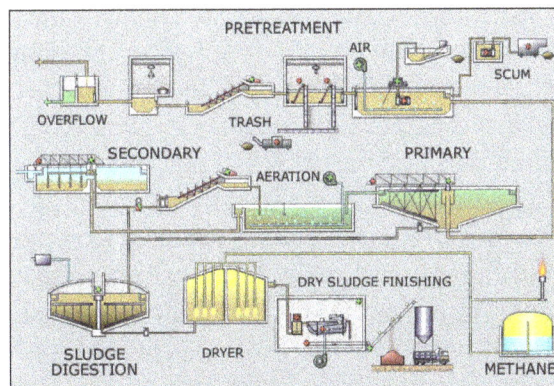

Simplified process flow diagram for a typical large-scale treatment plant.

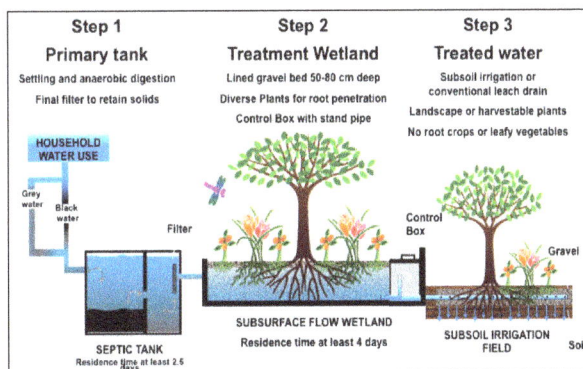

Process flow diagram for a typical treatment plant via subsurface flow constructed wetlands (SFCW).

Pretreatment

Pretreatment removes all materials that can be easily collected from the raw sewage before they damage or clog the pumps and sewage lines of primary treatment clarifiers. Objects commonly removed during pretreatment include trash, tree limbs, leaves, branches, and other large objects.

The influent in sewage water passes through a bar screen to remove all large objects like cans, rags, sticks, plastic packets etc. carried in the sewage stream. This is most commonly done with an automated mechanically raked bar screen in modern plants serving large populations, while in smaller or less modern plants, a manually cleaned screen may be used. The raking action of a mechanical bar screen is typically paced according to the accumulation on the bar screens and flow rate. The solids are collected and later disposed in a landfill, or incinerated. Bar screens or mesh screens of varying sizes may be used to optimize solids removal. If gross solids are not removed, they become entrained in pipes and moving parts of the treatment plant, and can cause substantial damage and inefficiency in the process.

Grit Removal

Grit consists of sand, gravel, cinders, and other heavy materials. It also includes organic matter such as eggshells, bone chips, seeds, and coffee grounds. Pretreatment may include a sand or grit channel or chamber, where the velocity of the incoming sewage is adjusted to allow the settlement of sand and grit. Grit removal is necessary to: (1) reduce formation of heavy deposits in aeration tanks, aerobic digesters, pipelines, channels, and conduits; (2) reduce the frequency of digester cleaning caused by excessive accumulations of grit; and (3) protect moving mechanical equipment from abrasion and accompanying abnormal wear. The removal of grit is essential for equipment with closely machined metal surfaces such as comminutors, fine screens, centrifuges, heat exchangers, and high pressure diaphragm pumps. Grit chambers come in 3 types: horizontal grit chambers, aerated grit chambers and vortex grit chambers. Vortex type grit chambers include mechanically induced vortex, hydraulically induced vortex, and multi-tray vortex separators. Given that traditionally, grit removal systems have been designed to remove clean inorganic particles that are greater than 0.210 millimetres (0.0083 in), most grit passes through the grit removal flows under normal conditions. During periods of high flow deposited grit is resuspended and the quantity of grit reaching the treatment plant increases substantially. It is, therefore important that the grit removal system not only operate efficiently during normal flow conditions but also under sustained peak flows when the greatest volume of grit reaches the plant.

Flow Equalization

Clarifiers and mechanized secondary treatment are more efficient under uniform flow conditions. Equalization basins may be used for temporary storage of diurnal or wet-weather flow peaks. Basins provide a place to temporarily hold incoming sewage during plant maintenance and a means of diluting and distributing batch discharges of toxic or high-strength waste which might otherwise inhibit biological secondary treatment (including portable toilet waste, vehicle holding tanks, and septic tank pumpers). Flow equalization basins require variable discharge control, typically include provisions for bypass and cleaning, and may also include aerators. Cleaning may be easier if the basin is downstream of screening and grit removal.

Fat and Grease Removal

In some larger plants, fat and grease are removed by passing the sewage through a small tank where skimmers collect the fat floating on the surface. Air blowers in the base of the tank may also be used to help recover the fat as a froth. Many plants, however, use primary clarifiers with mechanical surface skimmers for fat and grease removal.

Primary Treatment

Primary treatment tanks in Oregon, USA.

In the primary sedimentation stage, sewage flows through large tanks, commonly called "pre-settling basins", "primary sedimentation tanks" or "primary clarifiers". The tanks are used to settle sludge while grease and oils rise to the surface and are skimmed off. Primary settling tanks are usually equipped with mechanically driven scrapers that continually drive the collected sludge towards a hopper in the base of the tank where it is pumped to sludge treatment facilities. Grease and oil from the floating material can sometimes be recovered for saponification (soap making).

Secondary Treatment

Secondary treatment is designed to substantially degrade the biological content of the sewage which are derived from human waste, food waste, soaps and detergent. The majority of municipal plants treat the settled sewage liquor using aerobic biological processes. To be effective, the biota require both oxygen and food to live. The bacteria and protozoa consume biodegradable soluble organic contaminants (e.g. sugars, fats, organic short-chain carbon molecules, etc.) and bind much of the less soluble fractions into floc.

Secondary treatment systems are classified as fixed-film or suspended-growth systems:

- Fixed-film or attached growth systems include trickling filters, constructed wetlands, bio-towers, and rotating biological contactors, where the biomass grows on media and the sewage passes over its surface. The fixed-film principle has further developed into moving bed biofilm reactors (MBBR) and Integrated Fixed-Film Activated Sludge (IFAS) processes. An MBBR system typically requires a smaller footprint than suspended-growth systems.

- Suspended-growth systems include activated sludge, where the biomass is mixed with the sewage and can be operated in a smaller space than trickling filters that treat the same amount of water. However, fixed-film systems are more able to cope with drastic changes in the amount of biological material and can provide higher removal rates for organic material and suspended solids than suspended growth systems.

Secondary clarifier at a rural treatment plant.

Some secondary treatment methods include a secondary clarifier to settle out and separate biological floc or filter material grown in the secondary treatment bioreactor.

Tertiary Treatment

The purpose of tertiary treatment is to provide a final treatment stage to further improve the effluent quality before it is discharged to the receiving environment (sea, river, lake, wet lands, ground, etc.). More than one tertiary treatment process may be used at any treatment plant. If disinfection is practised, it is always the final process. It is also called "effluent polishing".

Filtration

Sand filtration removes much of the residual suspended matter. Filtration over activated carbon, also called carbon adsorption, removes residual toxins.

Lagoons or Ponds

Lagoons or ponds provide settlement and further biological improvement through storage in large man-made ponds or lagoons. These lagoons are highly aerobic and colonization by native macrophytes, especially reeds, is often encouraged. Small filter-feeding invertebrates such as *Daphnia* and species of *Rotifera* greatly assist in treatment by removing fine particulates.

A sewage treatment plant and lagoon in Everett, Washington, USA.

Biological Nutrient Removal

Biological nutrient removal (BNR) is regarded by some as a type of secondary treatment process, and by others as a tertiary (or "advanced") treatment process.

Wastewater may contain high levels of the nutrients nitrogen and phosphorus. Excessive release to the environment can lead to a buildup of nutrients, called eutrophication, which can in turn encourage the overgrowth of weeds, algae, and cyanobacteria (blue-green algae). This may cause an algal bloom, a rapid growth in the population of algae. The algae numbers are unsustainable and eventually most of them die. The decomposition of the algae by bacteria uses up so much of the oxygen in the water that most or all of the animals die, which creates more organic matter for the bacteria to decompose. In addition to causing deoxygenation, some algal species produce toxins that contaminate drinking water supplies. Different treatment processes are required to remove nitrogen and phosphorus.

Nitrogen Removal

Nitrogen is removed through the biological oxidation of nitrogen from ammonia to nitrate (nitrification), followed by denitrification, the reduction of nitrate to nitrogen gas. Nitrogen gas is released to the atmosphere and thus removed from the water.

Nitrification itself is a two-step aerobic process, each step facilitated by a different type of bacteria. The oxidation of ammonia (NH_3) to nitrite (NO_2^-) is most often facilitated by *Nitrosomonas* spp. ("nitroso" referring to the formation of a nitroso functional group). Nitrite oxidation to nitrate (NO_3^-), though traditionally believed to be facilitated by *Nitrobacter* spp. (nitro referring the formation of a nitro functional group), is now known to be facilitated in the environment almost exclusively by *Nitrospira* spp.

Denitrification requires anoxic conditions to encourage the appropriate biological communities to form. It is facilitated by a wide diversity of bacteria. Sand filters, lagooning and reed beds can all be used to reduce nitrogen, but the activated sludge process (if designed well) can do the job the most easily. Since denitrification is the reduction of nitrate to dinitrogen (molecular nitrogen) gas, an electron donor is needed. This can be, depending on the waste water, organic matter (from feces), sulfide, or an added donor like methanol. The sludge in the anoxic tanks (denitrification tanks) must be mixed well (mixture of recirculated mixed liquor, return activated sludge [RAS], and raw influent) e.g. by using submersible mixers in order to achieve the desired denitrification.

Sometimes the conversion of toxic ammonia to nitrate alone is referred to as tertiary treatment.

Over time, different treatment configurations have evolved as denitrification has become more sophisticated. An initial scheme, the Ludzack–Ettinger Process, placed an anoxic treatment zone before the aeration tank and clarifier, using the return activated sludge (RAS) from the clarifier as a nitrate source. Influent wastewater (either raw or as effluent from primary clarification) serves as the electron source for the facultative bacteria to metabolize carbon, using the inorganic nitrate as a source of oxygen instead of dissolved molecular oxygen. This denitrification scheme was naturally limited to the amount of soluble nitrate present in the RAS. Nitrate reduction was limited because RAS rate is limited by the performance of the clarifier.

The "Modified Ludzak–Ettinger Process" (MLE) is an improvement on the original concept, for it recycles mixed liquor from the discharge end of the aeration tank to the head of the anoxic tank to provide a consistent source of soluble nitrate for the facultative bacteria. In this instance, raw wastewater continues to provide the electron source, and sub-surface mixing maintains the bacteria in contact with both electron source and soluble nitrate in the absence of dissolved oxygen.

Many sewage treatment plants use centrifugal pumps to transfer the nitrified mixed liquor from the aeration zone to the anoxic zone for denitrification. These pumps are often referred to as *Internal Mixed Liquor Recycle* (IMLR) pumps. IMLR may be 200% to 400% the flow rate of influent wastewater (Q). This is in addition to Return Activated Sludge (RAS) from secondary clarifiers, which may be 100% of Q. (Therefore, the hydraulic capacity of the tanks in such a system should handle at least 400% of annual average design flow (AADF). At times, the raw or primary effluent wastewater must be carbon-supplemented by the addition of methanol, acetate, or simple food waste (molasses, whey, plant starch) to improve the treatment efficiency. These carbon additions should be accounted for in the design of a treatment facility's organic loading. Further modifications to the MLE were to come: Bardenpho and Biodenipho processes include additional anoxic and oxidative processes to further polish the conversion of nitrate ion to molecular nitrogen gas.

The use of an anaerobic tank following the initial anoxic process allows for luxury uptake of phosphorus by bacteria, thereby biologically reducing orthophosphate ion in the treated wastewater. Even newer improvements, such as Anammox Process, interrupt the formation of nitrate at the nitrite stage of nitrification, shunting nitrite-rich mixed liquor activated sludge to treatment where nitrite is then converted to molecular nitrogen gas, saving energy, alkalinity, and secondary carbon sourcing. Anammox (ANaerobic AMMonia OXidation) works by artificially extending detention time and preserving denitrifying bacteria through the use of substrate added to the mixed liquor and continuously recycled from it prior to secondary clarification. Many other proprietary schemes are being deployed, including DEMON, Sharon-ANAMMOX, ANITA-Mox, and DeAmmon. The bacteria Brocadia anammoxidans can remove ammonium from waste water through anaerobic oxidation of ammonium to hydrazine, a form of rocket fuel.

Phosphorus Removal

Every adult human excretes between 200 and 1,000 grams (7.1 and 35.3 oz) of phosphorus annually. Studies of United States sewage in the late 1960s estimated mean per capita contributions of 500 grams (18 oz) in urine and feces, 1,000 grams (35 oz) in synthetic detergents, and lesser variable amounts used as corrosion and scale control chemicals in water supplies. Source control via alternative detergent formulations has subsequently reduced the largest contribution, but the content of urine and feces will remain unchanged. Phosphorus removal is important as it is a limiting nutrient for algae growth in many fresh water systems. It is also particularly important for water reuse systems where high phosphorus concentrations may lead to fouling of downstream equipment such as reverse osmosis.

Phosphorus can be removed biologically in a process called enhanced biological phosphorus removal. In this process, specific bacteria, called polyphosphate-accumulating organisms (PAOs), are selectively enriched and accumulate large quantities of phosphorus within their cells (up to 20 percent of their mass). When the biomass enriched in these bacteria is separated from the treated water, these biosolids have a high fertilizer value.

Phosphorus removal can also be achieved by chemical precipitation, usually with salts of iron (e.g. ferric chloride), aluminum (e.g. alum), or lime. This may lead to excessive sludge production as hydroxides precipitate and the added chemicals can be expensive. Chemical phosphorus removal requires significantly smaller equipment footprint than biological removal, is easier to operate and is often more reliable than biological phosphorus removal. Another method for phosphorus removal is to use granular laterite.

Some systems use both biological phosphorus removal and chemical phosphorus removal. The chemical phosphorus removal in those systems may be used as a backup system, for use when the biological phosphorus removal is not removing enough phosphorus, or may be used continuously. In either case, using both biological and chemical phosphorus removal has the advantage of not increasing sludge production as much as chemical phosphorus removal on its own, with the disadvantage of the increased initial cost associated with installing two different systems.

Once removed, phosphorus, in the form of a phosphate-rich sewage sludge, may be dumped in a landfill or used as fertilizer. In the latter case, the treated sewage sludge is also sometimes referred to as biosolids.

Disinfection

The purpose of disinfection in the treatment of waste water is to substantially reduce the number of microorganisms in the water to be discharged back into the environment for the later use of drinking, bathing, irrigation, etc. The effectiveness of disinfection depends on the quality of the water being treated (e.g., cloudiness, pH, etc.), the type of disinfection being used, the disinfectant dosage (concentration and time), and other environmental variables. Cloudy water will be treated less successfully, since solid matter can shield organisms, especially from ultraviolet light or if contact times are low. Generally, short contact times, low doses and high flows all militate against effective disinfection. Common methods of disinfection include ozone, chlorine, ultraviolet light, or sodium hypochlorite. Chloramine, which is used for drinking water, is not used in the treatment of waste water because of its persistence. After multiple steps of disinfection, the treated water is ready to be released back into the water cycle by means of the nearest body of water or agriculture. Afterwards, the water can be transferred to reserves for everyday human uses.

Chlorination remains the most common form of waste water disinfection in North America due to its low cost and long-term history of effectiveness. One disadvantage is that chlorination of residual organic material can generate chlorinated-organic compounds that may be carcinogenic or harmful to the environment. Residual chlorine or chloramines may also be capable of chlorinating organic material in the natural aquatic environment. Further, because residual chlorine is toxic to aquatic species, the treated effluent must also be chemically dechlorinated, adding to the complexity and cost of treatment.

Ultraviolet (UV) light can be used instead of chlorine, iodine, or other chemicals. Because no chemicals are used, the treated water has no adverse effect on organisms that later consume it, as may be the case with other methods. UV radiation causes damage to the genetic structure of bacteria, viruses, and other pathogens, making them incapable of reproduction. The key disadvantages of UV disinfection are the need for frequent lamp maintenance and replacement and the need for a highly treated effluent to ensure that the target microorganisms are not shielded from the UV radiation (i.e., any solids present in the treated effluent may protect microorganisms from the UV light). In the United Kingdom, UV light is becoming the most common means of disinfection because of the concerns about the impacts of chlorine in chlorinating residual organics in the wastewater and in chlorinating organics in the receiving water. Some sewage treatment systems in Canada and the US also use UV light for their effluent water disinfection.

Ozone (O_3) is generated by passing oxygen (O_2) through a high voltage potential resulting in a third oxygen atom becoming attached and forming O_3. Ozone is very unstable and reactive and oxidizes most organic material it comes in contact with, thereby destroying many pathogenic microorganisms. Ozone is considered to be safer than chlorine because, unlike chlorine which has to be stored on site (highly poisonous in the event of an accidental release), ozone is generated on-site as needed from the oxygen in the ambient air. Ozonation also produces fewer disinfection by-products than chlorination. A disadvantage of ozone disinfection is the high cost of the ozone generation equipment and the requirements for special operators.

Fourth Treatment Stage

Micropollutants such as pharmaceuticals, ingredients of household chemicals, chemicals used in small businesses or industries, environmental persistent pharmaceutical pollutant (EPPP) or pesticides may not be eliminated in the conventional treatment process (primary, secondary and tertiary treatment) and therefore lead to water pollution. Although concentrations of those substances and their decomposition products are quite low, there is still a chance to harm aquatic organisms. For pharmaceuticals, the following substances have been identified as "toxicologically relevant": substances with endocrine disrupting effects, genotoxic substances and substances that enhance the development of bacterial resistances. They mainly belong to the group of environmental persistent pharmaceutical pollutants. Techniques for elimination of micropollutants via a fourth treatment stage during sewage treatment are implemented in Germany, Switzerland, Sweden and the Netherlands and tests are ongoing in several other countries. Such process steps mainly consist of activated carbon filters that adsorb the micropollutants. The combination of advanced oxidation with ozone followed by GAC, Granulated Activated Carbon, has been suggested as a cost-effective treatment combination for pharmaceutical residues. For a full reduction of microplasts the combination of ultra filtration followed by GAC has been suggested. Also the use of enzymes such as the enzyme laccase is under investigation. A new concept which could provide an energy-efficient treatment of micropollutants could be the use of laccase secreting fungi cultivated at a wastewater treatment plant to degrade micropollutants and at the same time to provide enzymes at a cathode of a microbial biofuel cells. Microbial biofuel cells are investigated for their property to treat organic matter in wastewater.

To reduce pharmaceuticals in water bodies, also "source control" measures are under investigation, such as innovations in drug development or more responsible handling of drugs.

Environment Aspects

Many processes in a wastewater treatment plant are designed to mimic the natural treatment processes that occur in the environment, whether that environment is a natural water body or the ground. If not overloaded, bacteria in the environment will consume organic contaminants, although this will reduce the levels of oxygen in the water and may significantly change the overall ecology of the receiving water. Native bacterial populations feed on the organic contaminants, and the numbers of disease-causing microorganisms are reduced by natural environmental conditions such as predation or exposure to ultraviolet radiation. Consequently, in cases where the receiving environment provides a high level of dilution, a high degree of wastewater treatment may not be required. However, recent evidence has demonstrated that very low levels of specific contaminants in wastewater, including hormones (from animal husbandry and residue from human hormonal contraception methods) and synthetic materials such as phthalates that mimic hormones in their action, can have an unpredictable adverse impact on the natural biota and potentially on humans if the water is re-used for drinking water. In the US and EU, uncontrolled discharges of wastewater to the environment are not permitted under law, and strict water quality requirements are to be met, as clean drinking water is essential. A significant threat in the coming decades will be the increasing uncontrolled discharges of wastewater within rapidly developing countries.

Treated water from WWTP Děčín, Czech Republic.

Treated water drained to the Elbe river, Děčín, Czech Republic.

Effects on Biology

Sewage treatment plants can have multiple effects on nutrient levels in the water that the treated sewage flows into. These nutrients can have large effects on the biological life in the water in contact with the effluent. Stabilization ponds (or sewage treatment ponds) can include any of the following:

- Oxidation ponds, which are aerobic bodies of water usually 1–2 metres (3 ft 3 in–6 ft 7 in) in depth that receive effluent from sedimentation tanks or other forms of primary treatment.

- Dominated by algae.

- Polishing ponds are similar to oxidation ponds but receive effluent from an oxidation pond or from a plant with an extended mechanical treatment.

- Dominated by zooplankton.

- Facultative lagoons, raw sewage lagoons, or sewage lagoons are ponds where sewage is added with no primary treatment other than coarse screening. These ponds provide effective treatment when the surface remains aerobic; although anaerobic conditions may develop near the layer of settled sludge on the bottom of the pond.

- Anaerobic lagoons are heavily loaded ponds.

- Dominated by bacteria.

- Sludge lagoons are aerobic ponds, usually 2 to 5 metres (6 ft 7 in to 16 ft 5 in) in depth, that receive anaerobically digested primary sludge, or activated secondary sludge under water.

- Upper layers are dominated by algae.

Phosphorus limitation is a possible result from sewage treatment and results in flagellate-dominated plankton, particularly in summer and fall.

A phytoplankton study found high nutrient concentrations linked to sewage effluents. High nutrient concentration leads to high chlorophyll a concentrations, which is a proxy for primary production in marine environments. High primary production means high phytoplankton populations and most likely high zooplankton populations, because zooplankton feed on phytoplankton. However, effluent released into marine systems also leads to greater population instability.

The planktonic trends of high populations close to input of treated sewage is contrasted by the bacterial trend. In a study of *Aeromonas* spp. in increasing distance from a wastewater source, greater change in seasonal cycles was found the furthest from the effluent. This trend is so strong that the furthest location studied actually had an inversion of the *Aeromonas* spp. cycle in comparison to that of fecal coliforms. Since there is a main pattern in the cycles that occurred simultaneously at all stations it indicates seasonal factors (temperature, solar radiation, phytoplankton) control of the bacterial population. The effluent dominant species changes from *Aeromonas caviae* in winter to *Aeromonas sobria* in the spring and fall while the inflow dominant species is *Aeromonas caviae*, which is constant throughout the seasons.

Reuse

With suitable technology, it is possible to reuse sewage effluent for drinking water, although this is usually only done in places with limited water supplies, such as Windhoek and Singapore.

In arid countries, treated wastewater is often used in agriculture. For example, in Israel, about 50 percent of agricultural water use (total use was one billion cubic metres (3.5×10^{10} cu ft) in 2008) is provided through reclaimed sewer water. Future plans call for increased use of treated sewer water as well as more desalination plants as part of water supply and sanitation in Israel.

Constructed wetlands fed by wastewater provide both treatment and habitats for flora and fauna. Another example for reuse combined with treatment of sewage are the East Kolkata Wetlands in India. These wetlands are used to treat Kolkata's sewage, and the nutrients contained in the wastewater sustain fish farms and agriculture.

RECLAIMED WATER

Reclaimed or recycled water is the process of converting wastewater into water that can be reused for other purposes. Reuse may include irrigation of gardens and agricultural fields or replenishing surface water and groundwater (i.e., groundwater recharge). Reused water may also be directed toward fulfilling certain needs in residences (e.g. toilet flushing), businesses, and industry, and could even be treated to reach drinking water standards. This last option is called either "direct potable reuse" or "indirect potable" reuse, depending on the approach used. Colloquially, the term "toilet to tap" also refers to potable reuse.

Sequence of reclamation from left: raw sewage, plant effluent, and finally reclaimed water.

Reclaiming water for reuse applications instead of using freshwater supplies can be a water-saving measure. When used water is eventually discharged back into natural water sources, it can still have benefits to ecosystems, improving streamflow, nourishing plant life and recharging aquifers, as part of the natural water cycle.

Wastewater reuse is a long-established practice used for irrigation, especially in arid countries. Reusing wastewater as part of sustainable water management allows water to remain as an alternative water source for human activities. This can reduce scarcity and alleviate pressures on groundwater and other natural water bodies.

Achieving more sustainable sanitation and wastewater management will require emphasis on actions linked to resource management, such as wastewater reuse or excreta reuse that will keep valuable resources available for productive uses. This in turn supports human wellbeing and broader sustainability.

Simply stated, reclaimed water is water that is used more than one time before it passes back into the natural water cycle. Advances in wastewater treatment technology allow communities to reuse water for many different purposes. The water is treated differently depending upon the source and use of the water and how it gets delivered.

Cycled repeatedly through the planetary hydrosphere, all water on Earth is recycled water, but the terms "recycled water" or "reclaimed water" typically mean wastewater sent from a home or business through a sewer system to a wastewater treatment plant, where it is treated to a level consistent with its intended use.

The World Health Organization has recognized the following principal driving forces for wastewater reuse:

1. Increasing water scarcity and stress,

2. Increasing populations and related food security issues,

3. Increasing environmental pollution from improper wastewater disposal,

4. Increasing recognition of the resource value of wastewater, excreta and greywater.

Water recycling and reuse is of increasing importance, not only in arid regions but also in cities and contaminated environments.

Already, the groundwater aquifers that are used by over half of the world population are being over-drafted. Reuse will continue to increase as the world's population becomes increasingly urbanized and concentrated near coastlines, where local freshwater supplies are limited or are available only with large capital expenditure. Large quantities of freshwater can be saved by wastewater reuse and recycling, reducing environmental pollution and improving carbon footprint. Reuse can be an alternative water supply option.

Types and Applications

Most of the uses of water reclamation are non potable uses such as washing cars, flushing toilets, cooling water for power plants, concrete mixing, artificial lakes, irrigation for golf courses and public parks, and for hydraulic fracturing. Where applicable, systems run a dual piping system to keep the recycled water separate from the potable water.

Categories of use	Uses
Urban uses	Irrigation of public parks, sporting facilities, private gardens, roadsides; Street cleaning; Fire protection systems; Vehicle washing; Toilet flushing; Air conditioners; Dust control.
Agricultural uses	Food crops not commercially processed; Food crops commercially processed; Pasture for milking animals; Fodder; Fibre; Seed crops; Ornamental flowers; Orchards; Hydroponic culture; Aquaculture; Greenhouses; Viticulture.
Industrial uses	Processing water; Cooling water; Recirculating cooling towers; Washdown water; Washing aggregate; Making concrete; Soil compaction; Dust control.
Recreational uses	Golf course irrigation; Recreational impoundments with/without public access (e.g. fishing, boating, bathing); Aesthetic impoundments without public access; Snowmaking.
Environmental uses	Aquifer recharge; Wetlands; Marshes; Stream augmentation; Wildlife habitat; Silviculture.
Potable uses	Aquifer recharge for drinking water use; Augmentation of surface drinking water supplies; Treatment until drinking water quality.

De Facto Wastewater Reuse

De facto, unacknowledged or unplanned potable reuse refers to a situation where reuse of treated wastewater is, in fact, practiced but is not officially recognized. For example, a wastewater treatment plant from one city may be discharging effluents to a river which is used as a drinking water supply for another city downstream.

Unplanned Indirect Potable Use has existed for a long time. Large towns on the River Thames upstream of London (Oxford, Reading, Swindon, Bracknell) discharge their treated sewage ("non-potable water") into the Thames, which supplies water to London downstream. In the United States, the Mississippi River serves as both the destination of sewage treatment plant effluent and the source of potable water.

Urban Reuse

- Unrestricted: The use of reclaimed water for non-potable applications in municipal settings, where public access is not restricted.

- Restricted: The use of reclaimed water for non-potable applications in municipal settings, where public access is controlled or restricted by physical or institutional barriers, such as fencing, advisory signage, or temporal access restriction.

Agricultural Reuse

There are benefits of using recycled water for irrigation, including the lower cost compared to some other sources and consistency of supply regardless of season, climatic conditions and associated water restrictions. When reclaimed water is used for irrigation in agriculture, the nutrient (nitrogen and phosphorus) content of the treated wastewater has the benefit of acting as a fertilizer. This can make the reuse of excreta contained in sewage attractive.

The irrigation water can be used in different ways on different crops:

- Food crops to be eaten raw: crops which are intended for human consumption to be eaten raw or unprocessed.

- Processed food crops: crops which are intended for human consumption not to be eaten raw but after treatment process (i.e. cooked, industrially processed).

- Non-food crops: crops which are not intended for human consumption (e.g. pastures, forage, fiber, ornamental, seed, forest and turf crops).

In developing countries, agriculture is increasingly using untreated wastewater for irrigation - often in an unsafe manner. Cities provide lucrative markets for fresh produce, so are attractive to farmers. However, because agriculture has to compete for increasingly scarce water resources with industry and municipal users, there is often no alternative for farmers but to use water polluted with urban waste directly to water their crops.

There can be significant health hazards related to using untreated wastewater in agriculture. Wastewater from cities can contain a mixture of chemical and biological pollutants. In low-income countries, there are often high levels of pathogens from excreta. In emerging nations, where industrial development is outpacing environmental regulation, there are increasing risks from inorganic and organic chemicals. The World Health Organization, in collaboration with the Food and Agriculture Organization of the United Nations (FAO) and the United Nations Environmental Program (UNEP), has developed guidelines for safe use of wastewater in 2006. These guidelines advocate a 'multiple-barrier' approach wastewater use, for example by encouraging farmers to adopt various risk-reducing behaviours. These include ceasing irrigation a few days before harvesting to allow pathogens to die off in the sunlight, applying water carefully so it does not contaminate leaves likely to be eaten raw, cleaning vegetables with disinfectant or allowing fecal sludge used in farming to dry before being used as a human manure.

Environmental Reuse

The use of reclaimed water to create, enhance, sustain, or augment water bodies including wetlands, aquatic habitats, or stream flow is called "environmental reuse". For example, constructed wetlands fed by wastewater provide both wastewater treatment and habitats for flora and fauna.

Industrial Reuse

The use of reclaimed water to recharge aquifers that are not used as a potable water source.

Planned Potable Reuse

Planned potable reuse is publicly acknowledged as an intentional project to recycle water for drinking water. There are two ways in which potable water can be delivered for reuse - "Indirect Potable Reuse" (IPR) and "Direct Potable Reuse". Both these forms of reuse are commonly involve a more formal public process and public consultation program than is the case with de facto or unacknowledged reuse. In 'indirect' potable reuse applications, the reclaimed wastewater is used directly or mixed with other sources.

Direct potable reuse is also called "toilet to tap".

Some water agencies reuse highly treated effluent from municipal wastewater or resource recovery plants as a reliable, drought proof source of drinking water. By using advanced purification processes, they produce water that meets all applicable drinking water standards. System reliability and frequent monitoring and testing are imperative to them meeting stringent controls.

The water needs of a community, water sources, public health regulations, costs, and the types of water infrastructure in place, such as distribution systems, man-made reservoirs, or natural groundwater basins, determine if and how reclaimed water can be part of the drinking water supply. Some communities reuse water to replenish groundwater basins. Others put it into surface water reservoirs. In these instances the reclaimed water is blended with other water supplies and sits in storage for a certain amount of time before it is drawn out and gets treated again at a water treatment or distribution system. In some communities, the reused water is put directly into pipelines that go to a water treatment plant or distribution system.

Modern technologies such as reverse osmosis and ultraviolet disinfection are commonly used when reclaimed water will be mixed with the drinking water supply.

Indirect Potable Reuse

Indirect potable reuse (IPR) means the water is delivered to the consumer indirectly. After it is purified, the reused water blends with other supplies and sits a while in some sort of storage, man-made or natural, before it gets delivered to a pipeline that leads to a water treatment plant or distribution system. That storage could be a groundwater basin or a surface water reservoir.

Some municipalities are using and others are investigating Indirect Potable Reuse (IPR) of reclaimed water. For example, reclaimed water may be pumped into (subsurface recharge) or percolated down to (surface recharge) groundwater aquifers, pumped out, treated again, and finally used as drinking water. This technique may also be referred to as groundwater recharging. This includes slow processes of further multiple purification steps via the layers of earth/sand (absorption) and microflora in the soil (biodegradation).

IPR or even unplanned potable use of reclaimed wastewater is used in many countries, where the latter is discharged into groundwater to hold back saline intrusion in coastal aquifers. IPR has generally included some type of environmental buffer, but conditions in certain areas have created an urgent need for more direct alternatives.

IPR occurs through the augmentation of drinking water supplies with urban wastewater treated to a level suitable for IPR followed by an environmental buffer (e.g. rivers, dams, aquifers, etc.) that precedes drinking water treatment. In this case, urban wastewater passes through a series of treatment steps that encompasses membrane filtration and separation processes (e.g. MF, UF and RO), followed by an advanced chemical oxidation process (e.g. UV, UV+H_2O_2, ozone).

Direct Potable Reuse

Direct potable reuse means the reused water is put directly into pipelines that go to a water treatment plant or distribution system. Direct potable reuse may occur with or without "engineered storage" such as underground or above ground tanks.

In a Direct Potable Reuse (DPR) scheme, water is put directly into pipelines that go to a water treatment plant or distribution system. Direct potable reuse may occur with or without "engineered storage" such as underground or above ground tanks. In other words, DPR is the introduction of reclaimed water derived from urban wastewater after extensive treatment and monitoring to assure that strict water quality requirements are met at all times, directly into a municipal water supply system.

Reuse in Space

Wastewater reclamation can be especially important in relation to human spaceflight. In 1998, NASA announced it had built a human waste reclamation bioreactor designed for use in the International Space Station and a manned Mars mission. Human urine and feces are input into one end of the reactor and pure oxygen, pure water, and compost (humanure) are output from the other end. The soil could be used for growing vegetables, and the bioreactor also produces electricity.

Aboard the International Space Station, astronauts have been able to drink recycled urine due to the introduction of the ECLSS system. The system costs $250 million and has been working since May 2009. The system recycles wastewater and urine back into potable water used for drinking, food preparation, and oxygen generation. This cuts back on the need for resupplying the space station so often.

Benefits

Water/wastewater reuse, as an alternative water source, can provide significant economic, social and environmental benefits, which are key motivators for implementing such reuse programmes. Specifically, in agriculture, irrigation with wastewater may contribute to improve production yields, reduce the ecological footprint and promote socioeconomic benefits. These benefits include:

- Increased water availability.

- Drinking water substitution - keep drinking water for drinking and reclaimed water for non-drinking use (i.e. industry, cleaning, irrigation, domestic uses, toilet flushing, etc.).

- Reduced over-abstraction of surface and groundwater.

- Reduced energy consumption associated with production, treatment, and distribution of water compared to using deep groundwater resources, water importation or desalination.

- Reduced nutrient loads to receiving waters (i.e. rivers, canals and other surface water resources).

- Reduced manufacturing costs of using high quality reclaimed water.

- Increased agricultural production (i.e. crop yields).

- Reduced application of fertilizers (i.e. conservation of nutrients, reducing the need for artificial fertilizer (e.g. soil nutrition by the nutrients existing in the treated effluents)).

- Enhanced environmental protection by restoration of streams, wetlands and ponds.

- Increased employment and local economy (e.g. tourism, agriculture).

Design Considerations

Distribution

A lavender-colored pipeline carrying nonpotable water in a dual piping system in Mountain View, California, U.S.

Nonpotable reclaimed water is often distributed with a dual piping network that keeps reclaimed water pipes completely separate from potable water pipes.

In many cities using reclaimed water, it is now in such demand that consumers are only allowed to use it on assigned days. Some cities that previously offered unlimited reclaimed water at a flat rate are now beginning to charge citizens by the amount they use.

Treatment Processes

For many types of reuse applications wastewater must pass through numerous sewage treatment process steps before it can be used. Steps might include screening, primary settling, biological treatment, tertiary treatment (for example reverse osmosis), and disinfection. It is possible to acquire nitrogen from sewage and produce ammonium nitrate. This generates a revenue, and produces a useful fertilizer for farmers.

There are several technologies used to treat wastewater for reuse. A combination of these technologies can meet strict treatment standards and make sure that the processed water is hygienically safe, meaning free from bacteria and viruses. The following are some of the typical technologies: Ozonation, ultrafiltration, aerobic treatment (membrane bioreactor), forward osmosis, reverse osmosis, advanced oxidation.

Wastewater is generally treated to only secondary level treatment when used for irrigation. A pump station distributes reclaimed water to users around the city. This may include golf courses, agricultural uses, cooling towers, or in land fills.

Alternative Options

Rather than treating wastewater for reuse purposes, other options can achieve similar effects of freshwater savings:

- Greywater reuse systems: At a household level, treated or untreated greywater may be used for flush toilets or to water the garden.

- Rainwater harvesting and stormwater recovery: Urban design systems which incorporate rainwater harvesting and reduce runoff are known as Water Sensitive Urban Design (WSUD) in Australia, Low Impact Development (LID) in the United States and Sustainable urban drainage systems (SUDS) in the United Kingdom.

- Seawater desalination: An energy-intensive process where salt and other minerals are removed from seawater to produce potable water for drinking and irrigation, typically through membrane filtration (reverse-osmosis), and steam-distillation.

Costs

The cost of reclaimed water exceeds that of potable water in many regions of the world, where a fresh water supply is plentiful. However, reclaimed water is usually sold to citizens at a cheaper rate to encourage its use. As fresh water supplies become limited from distribution costs, increased population demands, or climate change reducing sources, the cost ratios will evolve also. The evaluation of reclaimed water needs to consider the entire water supply system, as it may bring important value of flexibility into the overall system.

Reclaimed water systems usually require a dual piping network, often with additional storage tanks, which adds to the costs of the system.

Barriers to Implementation

- Full-scale implementation and operation of water reuse schemes still face regulatory, economic, social and institutional challenges.

- Economic viability of water reuse schemes.

- Costs of water quality monitoring and identification of contaminants. Difficulties in contaminant identification may include the separation of inorganic and organic pollutants, microorganisms, colloids, and others.

- Full cost recovery from water reuse schemes - lack of financial water pricing systems comparable to already subsidized conventional treatment plants.

- Psychological barriers, sometimes referred to as the "yuck factor" can also be an impediment to implementation, particularly for direct potable reuse plans. These psychological factors appear to be closely associated with disgust, specifically pathogen avoidance.

Environmental Aspects

There is debate about possible health and environmental effects. To address these concerns, A Risk Assessment Study of potential health risks of recycled water and comparisons to conventional Pharmaceuticals and Personal Care Product (PPCP) exposures was conducted by the WateReuse Research Foundation. For each of four scenarios in which people come into contact with recycled water used for irrigation - children on the playground, golfers, and landscape, and agricultural workers - the findings from the study indicate that it could take anywhere from a few years to millions of years of exposure to nonpotable recycled water to reach the same exposure to PPCPs that we get in a single day through routine activities.

Using reclaimed water for non-potable uses saves potable water for drinking, since less potable water will be used for non-potable uses.

It sometimes contains higher levels of nutrients such as nitrogen, phosphorus and oxygen which may somewhat help fertilize garden and agricultural plants when used for irrigation.

The usage of water reclamation decreases the pollution sent to sensitive environments. It can also enhance wetlands, which benefits the wildlife depending on that eco-system. It also helps to stop the chances of drought as recycling of water reduces the use of fresh water supply from underground sources. For instance, The San Jose/Santa Clara Water Pollution Control Plant instituted a water recycling program to protect the San Francisco Bay area's natural salt water marshes.

The main potential risks that are associated with reclaimed wastewater reuse for irrigation purposes, when the treatment is not adequate are the following:

- Contamination of the food chain with microcontaminants, pathogens (i.e. bacteria, viruses, protozoa, helminths), or antibiotic resistance determinants.

- Soil salinization and accumulation of various unknown constituents that might adversely affect agricultural production.

- Distribution of the indigenous soil microbial communities.

- Alteration of the physicochemical and microbiological properties of the soil and contribution to the accumulation of chemical/biological contaminants (e.g. heavy metals, chemicals (i.e. boron, nitrogen, phosphorus, chloride, sodium, pesticides/herbicides), natural chemicals (i.e. hormones), contaminants of emerging concern (CECs) (i.e. pharmaceuticals and their metabolites, personal care products, household chemicals and food additives and their transformation products) in it and subsequent uptake by plants and crops.

- Excessive growth of algae and vegetation in canals carrying wastewater (i.e. eutrophication).

- Groundwater quality degradation by the various reclaimed water contaminants, migrating and accumulating in the soil and aquifers.

CHALLENGE OF SUSTAINABLE WASTE MANAGEMENT

Waste is a global issue; if it is not properly dealt with, it poses a threat to public health and the environment. It is a growing issue linked directly to the way society produces and consumes. It concerns everyone. Waste management is one of the essential utility services underpinning society in the 21st Century, particularly in urban areas.

Waste management is a basic human need and can also be regarded as a basic human right. Ensuring proper sanitation and solid waste management ranks alongside the provision of potable water, shelter, food, energy, transport and communications; all are essential to society and to the economy as a whole. Despite this, the public and political profile of waste management is often lower than other utility services. Unfortunately, the consequences of doing little or even nothing to address waste management can be very costly to society and to the economy overall.

In the absence of waste regulations and their rigorous implementation and enforcement, generators of waste tend to opt for the cheapest available course of action. For example, household solid waste may be dumped in the street, on vacant land, or into drains, streams or other watercourses, or it may be burned to reduce the irritation of accumulating piles of waste.

By definition, uncontrolled waste is not managed and thus not measured, making it difficult to estimate the size of the problem and the scale of the associated costs. However, the evidence suggests that in a middle- or low-income city, the costs to society and the economy are about five to 10 times what sound solid waste management (SWM) would cost per capita. It is dramatically cheaper to manage waste now in an environmentally sound manner than to clean up in future years the "sins of the past".

Moving from Waste Management to Resource Management

Many developed countries have made great strides in addressing waste management, particularly since the environment came onto the international agenda in the 1960s, and there are many good practice examples available for the international community to learn from.

However, the initial focus was on waste after it had been discarded, whereas now attention has moved upstream, addressing the problem at its source through, for example, designing out waste, preventing its generation, reducing both the quantities and the uses of hazardous substances, minimising and reusing, and where residuals do occur, keeping them concentrated and separate to preserve their intrinsic value for recycling and recovery and preventing them from contaminating other waste that still has economic value for recovery.

The goal is to move the fundamental thinking away from "waste disposal" to "waste management" and from "waste" to "resources" — hence the updated terminology "waste and resource management" and "resource management", as part of the "circular economy". In this regard, the Global Waste Management Outlook interfaces with the earlier Global Outlook on Sustainable Consumption and Production policies.

Low- and middle-income countries still face major challenges in ensuring universal access to waste collection services, eliminating uncontrolled disposal and burning and moving towards environmentally sound management for all waste. Addressing these challenges is made even more difficult by forecasts that major cities in the lowest income countries are likely to double in population over the next 20 or so years, which is also likely to increase the local political priority given to waste issues. Low- and middle-income countries need to devise and implement innovative and effective policies and practices to promote waste prevention and stem the relentless increase in waste per capita as economies develop.

Waste Management as an Entry Point to Sustainable Development

Waste management is an issue that impacts many parts of society and the economy. It has strong linkages to a range of other global challenges such as health, climate change, poverty reduction, food and resource security and sustainable production and consumption. The political case for action is significantly strengthened when waste management is viewed as an entry point to address a range of such sustainable development issues, many of which are difficult to tackle.

Waste management is well embedded within the Sustainable Development Goals (SDGs), being included either explicitly or implicitly in more than half of the 17 goals. Thus a strong argument can be made for the strategic importance of improving waste management, insofar as actions here will contribute to progress towards a range of SDG targets. Setting and monitoring global targets for waste management will thus contribute significantly to attaining the SDGs.

Impact of Resource Recovery on Financing

Resource recovery (e.g. recycling, composting), if properly conceived and implemented can reduce the financial impact of waste collection and disposal services. For example, the separation of recyclable materials (such as paper, glass, metals, and plastics) at a source of generation leads to a reduction in the quantities of waste, which local governments otherwise have to transport and dispose of at a landfill.

In economically developing countries, the mixed municipal waste stream typically contains in the order of 20% to 30% (by weight) of potentially recyclable inorganic materials. As the economic status of a particular country improves, consumption patterns change, and an increase can be expected in the percentage of recyclable materials in the waste.

Thus, savings in disposal costs may be available in the future if additional quantities of recyclable materials are recovered and marketed. In addition, the segregation and processing of the organic matter in waste can make a sizeable contribution to the reduction of quantities requiring ultimate disposal, since organic matter typically constitutes 50% to 60% of the residential waste stream.

References

- Lienert, J.; Bürki, T.; Escher, B.I. (2007). "Reducing micropollutants with source control: Substance flow analysis of 212 pharmaceuticals in faeces and urine". Water Science & Technology. 56 (5): 87–96. Doi:10.2166/wst.2007.560. PMID 17881841

- Solid-waste-management, technology: britannica.com, Retrieved 7 July, 2019

- Sustainable-solid-waste-management: bioenergyconsult.com, Retrieved 9 january 2019

- "Sanitation Systems – Sanitation Technologies – Activated sludge". SSWM. 27 April 2018. Retrieved 31 October 2018

- Sustainable-practices-waste-management: conserve-energy-future.com, Retrieved 8 August, 2019

- "Water Reuse in Europe - Relevant guidelines, needs for and barriers to innovation". Retrieved 29 July 2016

- Solid-waste-management-and-sustainable-development: mg.co.za, Retrieved 10 February, 2019

INDEX

www.ingramcontent.com/pod-product-compliance
Lightning Source LLC
Chambersburg PA
CBHW082044190326

41458CB00010B/3453